history

DISCONTENT AT THE POLLS

DISCONTENT AT THE POLLS

A Study of Farmer and Labor Parties
1827-1948

BY MURRAY S. STEDMAN, JR.

AND SUSAN W. STEDMAN

NEW YORK

COLUMBIA UNIVERSITY PRESS 1950

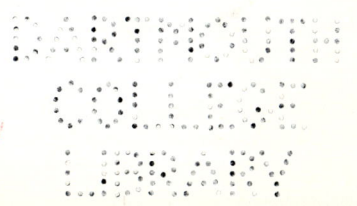

COPYRIGHT 1950 COLUMBIA UNIVERSITY PRESS, NEW YORK

PUBLISHED IN GREAT BRITAIN, CANADA, AND INDIA BY
GEOFFREY CUMBERLEGE, OXFORD UNIVERSITY PRESS
LONDON, TORONTO, AND BOMBAY

MANUFACTURED IN THE UNITED STATES OF AMERICA

PREFACE

THE THESIS OF THIS BOOK is that farmer and labor parties in the United States have performed two great functions. These functions are, first, to act as vehicles for the expression of political discontent, and secondly, to popularize issues which the major parties have at first ignored but later have adopted.

Precisely how these functions have been performed, and under what economic, geographical, and psychological conditions, this work endeavors to explain.

For having read the original manuscript and for having contributed valuable suggestions, we wish to acknowledge our indebtedness to Professors V. O. Key, Jr., of Johns Hopkins University, Elmer E. Schattschneider of Wesleyan University, Philip Taft of Brown University, and Lawrence H. Chamberlain of Columbia University.

Special thanks go to Professor Vincent H. Whitney of Brown University for his careful reading of the manuscript and for his numerous constructive comments. We are grateful to Professor Matthew C. Mitchell of Brown University for his continued encouragement.

We also wish to thank Dr. Louis H. Bean of the Department of Agriculture. Both his comments on our study and the insight contained in his various works on the American political process have been of great value.

We are indebted to *Fortune* magazine for special permission to use certain material in the *Fortune Survey*. Mr. Elmo Roper and members of his staff helped us by analyzing our use of the survey data.

In addition, we wish to thank Mr. Don C. Thorndike of Providence for his assistance in preparing the charts.

Since the statistics and other material were collected and the judgments were made solely by the authors, we can absolve all but ourselves for any errors of fact or interpretation.

MSS
SWS

Brown University
Providence, Rhode Island
July, 1949

CONTENTS

Introduction   3

1. The Parties: a Survey   6
2. Platforms: the Genealogy of Ideas   12
3. The Record at the Polls   32
4. Geographic Patterns of Protest Voting in Presidential Elections   50
5. The Economics of Protest Voting   78
6. Strategy and Tactics   102
7. Legal Barriers   125
8. The Components of Protest Voting: Economic Discontent   137
9. The Components of Protest Voting: Political Discontent   155
10. Perspective   168

Appendix A: Voting for Head of Tickets of Farmer and Labor Parties in Presidential and State-wide Elections, 1872–1948   173

Appendix B: Parties and States of Farmer and Labor Party Representatives in Congress, 1860–1948   175

Index   179

CHARTS

I. Vote for Farmer and Labor Parties in Presidential Elections, 1872–1948   35

II. Vote for Farmer and Labor Parties in Presidential and State-wide Elections, 1872–1948   38

III. Number of Candidates of Farmer and Labor Parties Elected to the United States House of Representatives, 1870–1948   45

IV. States with High Per Capita Voting for Farmer and Labor Parties in Presidential Elections, 1872–1948   55

V. States with High but Declining Per Capita Voting for Farmer and Labor Parties in Presidential Elections   58

VI. States with Increasing Per Capita Voting for Farmer and Labor Parties in Presidential Elections   62

VII. States with Moderate but Consistent Voting for Farmer and Labor Parties in Presidential Elections   67

VIII. States with Little or No Voting for Farmer and Labor Parties in Presidential Elections   72

IX. Voting for Head of Farmer and Labor Tickets in Presidential and State-wide Elections Compared with Wholesale Prices   79

X. Number of Farmer and Labor Candidates Elected to the U.S. House of Representatives and Wholesale Prices of Farm Products, 1860–1948   81

XI. Voting for Head of Farmer and Labor Tickets in Presidential and State-wide Elections Compared with the Ratio of Farm to Non-Farm Prices, 1870–1948 84

XII. Voting for Head of Farmer and Labor Tickets in Presidential and State-wide Elections Compared with Business Activity, 1872–1948 90

TABLES

1. Vote for Farmer and Labor Parties in Presidential Elections, 1872–1948 34

2. Economic Class and Opinion on the Role of Government in Economic Affairs 141

3. Occupation and Opinion on the Role of Government in Economic Affairs 145

DISCONTENT AT THE POLLS

INTRODUCTION

THIRD PARTIES are almost as native to the American political landscape as party conventions, smoke-filled rooms, and flowery campaign oratory.

The tradition of third parties goes back to the early years of the Republic. Among the notable elements of this tradition are farmer and labor parties—in many cases as American as apple pie and ice cream.

The effectiveness of farmer and labor parties is somewhat dimmed by the fact that the United States has a workable two-party system. The strength, the prestige, the resources of the two major parties have always obscured many of the contributions of lesser parties.

For the existence of a two-party system there are many reasons—habit, inertia, the single-member election district, the residence rule, the manner in which we choose our Presidents. Yet the fact remains that we have never had a pure and undiluted two-party system.[1] We have had scores of third parties, including many farmer-labor parties. It is likely that such parties will be on the scene so long as our political democracy lasts.

It is necessary in the interest of clarity to distinguish sharply between the two main categories of third parties. First, there is the third party which is really a secession from a major party. The Liberal Republicans of 1872, the Silver Republicans of 1896, and the Progressives of 1912 are all examples of secessions from a major party.

[1] One of the first writers to direct attention to this fact was John D. Hicks, in his article "The Third Party Tradition in American Politics," *Mississippi Valley Historical Review*, XX, No. 1 (June, 1933), 3–28.

The second general category of third parties includes all others. These may in turn be subdivided for the sake of convenience and analysis into farmer and labor parties and into agitational or educational parties. While the categories overlap, they are useful in order to illustrate various distinctions among the parties.

The Prohibition party is probably the leading historic example of an agitational organization. The Prohibitionists rarely allowed themselves to be distracted from the main issue—the liquor question. At no time did the members of the party deceive themselves into believing that they were seriously competing to win national elections. On occasion, however, they enjoyed local successes.

The other category of third parties is the one in which we are chiefly interested. These are the third parties which called themselves farmers' or workers' parties or which tried to attract their principal support from farmers or workers. In many cases these farmer-labor parties degenerated into mere agitational societies, as many feel to be the case today with parties of the extreme left. But in other instances the stakes were far higher, and farmer-labor parties competed seriously for state, regional, and even national control.

It is the farmer and labor parties that competed seriously to win elections which form the bulk of the subject matter of this study. One yardstick is particularly useful in measuring the efficacy of such parties. This is the election itself—the total number of votes accumulated by each candidate and each party, and the number of successful candidacies. In a political democracy such as the United States, this is a reasonable method of measurement.

Since, however, farmer and labor parties find themselves at a great disadvantage in competing with the major parties, an additional yardstick is also useful. This is an historical test. To what extent have the issues formulated by protest parties of the economically discontented eventually become the law of the land? While such a test is less precise than the mere measuring of votes, it tends to place the record of farmer and labor parties in a proper historical perspective.

From an immediate point of view, class politics—which farmer and labor parties must to some extent play—is not a paying proposition.[2] The proof of this observation is to be seen in the fact that professional politicians in America rarely find it expedient to operate along class lines. Rather, the middle class point of view, which is a nearly all-embracing point of view, is that adopted by state and city bosses and leaders.

In most parts of the United States it is necessary for a farmer or labor party in order to win an election to have the overwhelming support of an entire economic or social class. In practice such support is not often obtained. What Professor E. E. Schattschneider has called the "law of the imperfect mobilization of political interests" comes into play.

Yet, despite the many obstacles in their way, farmer and labor parties by and large have an impressive record. To show what this record is, including the factors making for successes and failures, is the primary purpose of this study. Such a purpose raises questions of the relationship of the protest vote to economic conditions, the regional characteristics of this vote, the political strategy and tactics used by the parties concerned, and the legal and psychological barriers which the parties encounter.

As the first step in assessing the significance of farmer and labor parties in our history, the parties themselves are introduced.

[2] For an interesting discussion of this point, see E. E. Schattschneider, *Party Government* (New York: Farrar, 1942), especially pp. 33, 85–93.

1

THE PARTIES: A SURVEY

THE PARTIES CONSIDERED in this study go back more than a hundred years in our history. To American politics they have contributed not only local color and material for specialized monographs; they have also contributed ideas which in many instances have formed the basis of subsequent legislation.[1]

The first group of labor parties sprang up in the late 1820's and early 1830's. Calling themselves Workingmen's parties, they were especially active in New York City, Philadelphia, Delaware, and in various heavily populated areas of New England. Most of them did not survive the panic of 1837. Probably the most widely known was the Equal Rights party (the Loco-Focos), which challenged Tammany in bitter battles for control of New York City from 1835 to 1837.

After the panic of 1837 little farmer or labor activity along political lines occurred until the post-Civil War period. Local labor parties were organized in the late 1860's, of which the Massachusetts Labor Reform party was typical and among the best known.

Not until 1872 did a labor party appear on a national scale. Called the Labor Reform party, this group represents the first attempt of farmers and workers to compete seriously in a Presidential election. Since the initial pioneering effort of the Labor Reform party there has been continuous representation of farmer and labor parties in Presidential elections.

---

[1] The labyrinth of leftist party politics is both involved and unrewarding; most of the "official" histories contain distortions and exaggerations which render their usefulness limited and their reliability questionable. Farmers' movements have fared better. Solon J. Buck's *The Agrarian Crusade* (New Haven: Yale University Press, 1920) and John D. Hicks' *The Populist Revolt* (Minneapolis: University of Minnesota Press, 1931) have set standards which leftist historians have thus far not equaled.

THE PARTIES 7

Examination of the voting record reveals that at no time in subsequent Presidential elections have farmer and labor parties been represented in less than twenty-three states.

At about the same time that the Labor Reform party was started, farmers' parties were beginning to establish themselves on a local scale throughout the West, under such names as the Independent party, the Reform party, the Anti-Monopoly party. From a national Independent party founded in Indianapolis in November, 1874, grew the Greenback party. It was supported by the national Anti-Monopoly party in the elections of 1884, and in 1887 and 1888 much of the Greenback membership passed into the newly organized Union Labor party.

It is of interest to note that at this same time state and city labor parties were coming into being in such widely separated areas as Chicago and New York. In the latter city the United Labor party ran Henry George for mayor in 1886. When George abandoned the party two years later, he contributed to its collapse.

The Granger-Greenback movement, with two decades of political experience behind it, turned into the Populist movement during the 1890's. The People's party, or Populists, were organized on a formal basis in 1890. Capitalizing on agrarian discontent they made a strong showing with their own candidate in 1892. In 1896 they endorsed the "boy orator of the Platte," the Democratic nominee William Jennings Bryan. By 1908 the Populist party was nearly dead; the official demise came in 1912 when only one hundred delegates attended the St. Louis convention held in August. Discouraged, the delegates made no nominations for President or Vice-President, but simply disbanded.

Venerable among any group of parties of the Western World is the Socialist Labor party. It was organized in 1876, largely by émigré German Marxists. In the Presidential election of 1892 it received 21,000 votes; in that of 1944 it received 45,000 votes. The party has approached modern industrial problems from the point of view of Marxism. For some years its most dynamic leader was Daniel De Leon, a life-long enemy of the American Federation of Labor's Samuel Gompers.

In 1898 the Social-Democratic party of America was established. During July, 1901, most socialists except those of the De Leon group merged into the Socialist party of America, under the leadership of Eugene V. Debs. Debs had announced his conversion to socialism after the Bryan campaign and eventually became the grand old man of the party. His reputation for humanitarianism and social justice extended far beyond the membership of the party. Thousands of ex-Populists, left without a home after 1900, readily gave their support to Debs and socialism.

After the First World War the Socialist party expelled its left-wing members, who promptly established a Communist group which soon split into many factions. In recent years there have been several secessions from the Communist party. The Socialist party suffered further internal difficulties in the mid-1930's and is now only a ghost of the Debs party which received more than 900,000 votes in the 1920 Presidential election.

The year 1924 saw the sudden birth, fruition, and collapse of a national farmer-labor party of sizable proportions. Supported by the railroad brotherhoods, the American Federation of Labor executive council, the Farmer-Labor party, and various other labor and some farm organizations, the Progressives of Senator LaFollette captured nearly five million votes in the Presidential contest. This amounted to some 16 per cent of the total vote cast, which compared favorably to the Democratic total of 28 per cent. With the death of their principal leader in 1925, the Progressives disappeared from

THE PARTIES

the political landscape. Even earlier, the withdrawal of labor groups foreshadowed the quick demise of the party.

From 1920 to 1932 a Farmer-Labor party appeared in national elections every year except 1924, when it endorsed Senator LaFollette. But the most significant group to call itself Farmer-Labor was the Minnesota organization, founded in the early 1920's as an outgrowth of the activities of the Farmers' Non-Partisan League in the Northwest. From 1931 to 1939 the party controlled much of the political life of Minnesota and in many ways was a very significant development in modern political behavior.

Another state party of great importance was the Wisconsin Progressive party. Controlled largely by the LaFollettes, this party represented a revolt against the state Republican machine. It lasted from 1934 until 1946.

In 1936 the American Labor party of New York State was formed under the inspiration of Labor's Nonpartisan League (dominated by CIO officials), some Old Guard Socialists, and various labor and liberal organizations. The right wing of the party seceded in 1944 to form the Liberal party. Early in 1948 the strongest single union to give its support to the ALP—the Amalgamated Clothing Workers—withdrew its support over the question of endorsing the candidacy of Henry A. Wallace for President.

Included in the general discussion and analysis of this volume is the record of the Union party of 1936. It presented a farmer-labor platform and called itself a farmer-labor party. With the discrediting of its leaders, the Rev. Charles Coughlin, the Rev. Gerald Smith, and Dr. Francis Townsend, the party disappeared immediately following Franklin Roosevelt's sensational victory in 1936. It should be noted that the Union party showed symptoms of incipient fascism and that its demise was generally applauded.

The most recent national party considered in this study is

the Progressive party formed in July, 1948. Headed by Henry A. Wallace and Glen Taylor, the party tried hard to win support from labor and farm organizations. Notwithstanding the fact that the new party received few endorsements from such groups, it thought of itself as a farmer and labor party. The record of the Progressive party at the polls in the 1948 Presidential elections is therefore shown in subsequent pages.

Also included in the general discussion are references to three "non-partisan" organizations which have issued platforms, encouraged independent candidacies, endorsed slates in the primaries, and sometimes acted like political parties. These organizations are the Farmers' Non-Partisan League, founded in 1915, the Political Action Committee of the Congress of Industrial Organizations, founded in 1943, and Labor's League for Political Education, established by the American Federation of Labor prior to the 1948 campaign.

Omitted from the discussion and analysis is the Progressive movement of 1912. It is clear that many former Populists, Socialists, and others who would normally have supported certain minor parties voted in 1912 for Theodore Roosevelt. Yet the reason for this deliberate omission is by now evident; the principal Bull Mooser himself and most of the local machines supporting him represented a temporary secession from the Republican party. Theodore Roosevelt was not running on an avowed farmer and labor ticket and was certainly not considered to be a farmer or labor candidate. We have excluded both Roosevelts and their parties from direct consideration for the same reason. They did not run on farmer or labor tickets, nor under farmer and labor labels. They both, of course, received millions of votes cast by farmers and workers as well as by other groups.

The parties and groups whose histories have thus been briefly sketched form the subject matter of this book. Beginning with the Workingmen's parties of the 1820's and

THE PARTIES 11

continuing through the Progressives of 1948, they have been an integral feature of the American political scene.

It is now appropriate to measure the effectiveness of these parties by one of the yardsticks of a political democracy. To what extent have their demands become law?

2

PLATFORMS: THE GENEALOGY OF IDEAS

HISTORICALLY, a vote for a farmer or labor party has not been a wasted ballot. This can be seen in the remarkable resemblance between platforms of the various farmer and labor parties in the past and many of the present public laws of the United States.

A few significant among many possible illustrations will be considered in order to emphasize the point that all ideas have their genealogy, and that many of our present laws are traceable in some measure to the agitation in former years by farmer and labor minority parties.

EARLY WORKINGMEN'S PARTIES

In the late 1820's the labor union movement began to mature along the Eastern seaboard. Out of a building-trades strike for a ten-hour day in Philadelphia in 1827 grew a "workingmen's party." This party put forth as its principal demand a proposal for a system of free, tax-supported schools. Other specific demands included the abolition or at least control of chartered monopolies, lottery control, prohibition, changes in the operation of the lien law, the insolvent law, and the militia law; and various tariff, money, and prison reforms.

At about the same time a similar workingmen's group came into existence in New York City, where the ten-hour day was already a reality, although its continuance was under challenge. This party was formed in 1829 under the leadership of Thomas Skidmore. Skidmore drew a sharp distinction between the rich and poor and called for a distribution of property so that each citizen could possess the rights which "belonged to him in a state of nature." When Skidmore and his followers seceded from the party, emphasis was then placed upon a plan of "state guardianship" by which children would be educated at state-supported boarding schools. This program received the support of Frances Wright and of Robert Dale Owen.

The New York City party, together with its branches upstate, denounced monopolies in general and various banking practices. Similar parties with like platforms sprang up in Wilmington, Delaware, and in Boston and many other of the principal New England cities.

While these early workingmen's parties did not survive the panic of 1837, several definite accomplishments may be traced directly or indirectly to their political activity. New York State abolished imprisonment for debt in 1831 and enacted a partial mechanics' lien law in 1832. The movement for free education received a tremendous boost from the workingmen's parties, particularly in Pennsylvania.[1]

THE GRANGER MOVEMENT

The Granger movement left its imprint on the laws of many of the older states of the Middle West. Complaints by farmers

---

[1] The general background for the discussion on the early workingmen's parties is taken from John R. Commons and associates in their *History of Labour in the United States* (4 vols., New York: Macmillan, 1918–1935). General background of the Granger and Populist movements is taken from Solon J. Buck, *The Agrarian Crusade* (New Haven: Yale University Press, 1920), John D. Hicks, *The Populist Revolt* (Minneapolis: University of Minnesota Press, 1931), and Fred E. Haynes, *Third Party Movements since the Civil War* (Iowa City: State Historical Society of Iowa, 1916). Material on the LaFollette movement may be found in Kenneth C. MacKay, *The Progressive Movement of 1924* (New York: Columbia University Press, 1947). The standard history of the Socialist movement up to 1928 is Nathan Fine's *Labor and Farmer Parties in the United States, 1828–1928* (New York: Rand School of Social Science, 1928). A recent short history of third parties—strongly colored with an agrarian bias—has been written by William B. Hesseltine, *The Rise and Fall of Third Parties* (Washington: Public Affairs Press, 1948). The above list merely touches the soil of literature on specific farmer and labor parties.

against what were considered to be exorbitant railroad rates and excessive warehouse charges led to demands for state regulation. The deflation of the postwar years, followed subsequently by the panic of 1873, served to intensify these demands.

An illustration is furnished by the Grangers in Illinois. The Illinois constitution of 1870 contained a clause which directed the legislature to correct railroad abuses. As a result, the legislature established a Railway and Warehouse Commission to regulate railroads, grain elevators, and warehouses. Iowa, Minnesota, and Wisconsin shortly followed suit.

When the validity of this legislation was challenged in the courts, it was eventually upheld by the United States Supreme Court in the so-called Granger cases. In a decision of enormous importance Chief Justice Waite in the case of *Munn v. Illinois* (94 U.S. 113; 1876) upheld an Illinois law regulating the charges of grain elevators. The Chief Justice enunciated the doctrine that private property affected with a public interest may be regulated in the public interest. On the same day the court sustained three Granger laws which established maximum passenger and freight rates.

A decade later, a more conservative Supreme Court held invalid rate regulation by a state legislative commission. Thereupon Congress accepted this challenge by passing the Interstate Commerce Act of 1887. To a very great extent federal regulation of railroads through the Interstate Commerce Commission owes its origin to the Granger demands that railroads should be regulated by government. When state regulation failed, the issue was put squarely up to Congress.

GREENBACKISM

The Greenback movement is best known for its inflationary demands. Yet other secondary demands deserve to be remembered, since many of them long outlived the greenback issue itself. The first attempt to organize political activity for Greenbackism came from the National Labor Union in 1868. At first the independent parties of the Granger period took little interest in the question; but the currency contraction which followed the panic of 1873 made the issue real on the Western farms.

The Labor Reform party of 1872 favored inflation and a number of other proposals. Among them was a resolution calling for a stop to importation of Chinese laborers. Another resolution demanded that states and municipalities employ labor on the basis of an eight-hour day, as the federal government had already begun to do.

An intimation of the Interstate Commerce Act of 1887 could be seen from the following demand:

Resolved, That it is the duty of the government to so exercise its power over railroads and telegraph corporations that they shall not in any case be privileged to exact such rates of freight, transportation or charges by whatever name, as may bear unduly or inequitably upon either producer or consumer.[2]

Both of the major parties included in their platforms references to the desirability of improving the civil service. But the strong stand of the Labor Reform party was perhaps the most sincere and certainly the most interesting of the 1872 campaign. The resolution read:

Resolved, That there should be such a reform in the Civil Service of the National Government as will remove it beyond all partisan influence, and place it in the charge and under the direction of intelligent and competent business men.[3]

It should not be supposed, however, that the genesis of civil service reform may be found exclusively or even principally in the platform of any single party, major or minor. Even

---

[2] The most convenient compilation of party platforms is to be found in the collection by Kirk H. Porter, *National Party Platforms* (New York: Macmillan, 1924). The citation is from p. 76. This and the following quotations are reprinted by permission of Professor Porter.
[3] *Ibid.*

before the Civil War some efforts were made in the direction of reform. Following the Civil War the lead was taken by brilliant reformers such as George W. Curtis, Dorman B. Eaton, Carl Schurz, Everett P. Wheeler, Charles J. Bonaparte, Richard Henry Dana, and William D. Foulke. Presidents Grant and Hayes made attempts to put merit principles into the operation of civil service. But perhaps the main impetus to reform came originally from the New York Civil Service Reform Association, formed in 1877, and later from the National Civil Service Reform League, organized in 1881.

The Labor Reform party contributed, as did other organizations, to the formation of public opinion favorable to civil service reform. Following the assassination of President Garfield by a disappointed office-seeker, Congress passed the Pendleton Act of 1883, and the Civil Service Commission was established. Similar laws were passed in New York State in 1883 and in Massachusetts in 1885.

At the Indianapolis convention of May 17, 1876, the number one demand of the Greenback party was for the immediate and unconditional repeal of the Specie Resumption Act of January 14, 1875. Four years later the same party, known variously as the Greenback Labor party and as the National party, adopted a much broader platform. Monopolies were again attacked, the paper tender issue was pressed, demands for government regulation of railroads and civil service reform were reiterated. Two additional and new proposals were presented:

All property should bear its just proportion of taxation, and we demand a graduated income tax.[4]

Resolved: That every citizen of due age, sound mind, and not a felon, be fully enfranchised and that this resolution be referred to the States, with recommendation for their favorable consideration.[5]

[4] Porter, *op. cit.*, p. 103. [5] *Ibid.*, p. 104.

The income tax was firmly established as a result of the 16th Amendment to the Constitution; it was proposed in 1909 and was adopted in 1913. The drive toward universal suffrage was successfully completed with the ratification of the 19th or "Susan B. Anthony" Amendment, which provided for woman suffrage, in 1920.

The Greenback party at a conference in Toledo in February of 1878 had also gone on record as favoring many labor objectives. It was at this conference, attended by representatives of farmer and labor organizations, that the Greenback Labor, or National party, was established. The Toledo resolutions gave recognition to labor's claims in that the party specifically called for a legislative reduction of the hours of labor, the establishment of labor bureaus, and the abolition of the contract system of employing prison labor.[6] The suppression of Chinese immigration—a strong demand of the unions—was also proposed.

In 1884, as in 1880, the Greenbackers again placed chief emphasis upon the money issue. In this they were joined by the Anti-Monopoly party, which also vigorously denounced alleged monopolies in transportation, money, and the communications industry. The Greenback National platform added two new demands which were prophetic of things to come:

We demand the amelioration of the condition of labor by enforcing the sanitary laws in industrial establishments, by the abolition of the convict labor system, by a rigid inspection of mines and factories, by a reduction of the hours of labor in industrial establishments, by fostering educational institutions, and by abolishing child labor.[7]

For the purpose of testing the sense of the people upon the subject, we are in favor of submitting to a vote of the people an amendment to the Constitution in favor of suffrage regardless of sex, and also on the subject of the liquor traffic.[8]

[6] Buck, *op. cit.*, p. 89.   [7] Porter, *op. cit.*, p. 126.
[8] *Ibid.*, p. 127.

The Union Labor platform of 1888, operating on the theory that the government should repeal certain "class legislation" in the interests of the ordinary worker and farmer, presented demands which have a modern ring. The party called for more housing, government ownership of communications and transportation, a freer (more flexible) money system, and arbitration to replace strikes in industrial disputes. Dissatisfaction with the Senate, which was termed a "millionaire's club," was expressed in the following resolution: "We demand a constitutional amendment making the United States senators elective by a direct vote of the people." [9]

The 17th Amendment, providing for the direct election of Senators, became effective in 1913.

THE POPULISTS

Many significant additions to our public laws since the turn of the century may be traced to the influence of the People's party, or Populists. The agitation was an outgrowth of earlier Granger and Greenback demands, but the scope of interest was far wider. By 1889 the resolutions adopted by the Northern and Southern Alliances were beginning to sound like party platforms. At the St. Louis convention of 1889 the Southern Alliance and the Knights of Labor agreed to act in unison in their legislative efforts before Congress.[10]

Stress was placed on monetary proposals, mostly of an inflationary nature. Considerable attention was also directed to the desirability of equitable taxation for all classes, economies in government spending, and government ownership of transportation and communications. To these demands the Northern Alliance added proposals calling for a graded income tax, a reduction of the tariff, improvements in the public school systems, and the acceptance of the Australian ballot.

It was at the St. Louis convention that President Macune

[9] Porter, *op. cit.*, p. 155. [10] The discussion is based mainly on Hicks, *op. cit.*

of the Southern Alliance presented his "sub-treasury" scheme. The essence of Macune's plan was for the government to abandon the practice of using certain banks as depositories. Instead, Macune proposed that a sub-treasury office be established in each county which produced for sale during one year farm products valued at half a million dollars. In connection with this office the government was to maintain warehouses or elevators in which the farmers might deposit their crops. In return, the government would give each depositor a certificate showing the amount and quality of the deposit, and also a loan and United States legal paper tender equal to 80 per cent of the current value of the products deposited. Interest on the loan was to be one per cent per year. If the farmer or person to whom he might sell his certificate did not redeem the property after one year, the government would sell the products at public auction. President Macune felt that his sub-treasury plan would increase the amount of currency in circulation, make the money more flexible, and permit farmers to hold crops for price rises.

At a convention held at Ocala, Florida, in December, 1890, the same groups demanded a Constitutional amendment providing for the direct election of United States Senators, and at a convention held in Omaha in January of the next year urged the direct election of the President and Vice-President.

The first convention of the People's party at Omaha in July, 1892, reiterated the bulk of these demands, emphasized the "free and unlimited coinage of silver and gold at the present legal ratio of sixteen to one," and proposed "that postal savings banks be established by the government for the safe deposit of the earnings of the people and to facilitate exchange."[11] The Omaha Resolutions also commended "to the favorable consideration of the people and the reform press the legisla-

[11] As quoted by Hicks, *op. cit.*, p. 443.

tive system known as the initiative and referendum."[12]

By 1896 the silver issue had been made paramount by the Populists, who went down to defeat with Bryan, fusion, and silver in one of the most bitterly contested Presidential elections of our history. What was left of the party had split into two groups by 1900. The Fusion faction cooperated with the Democrats, while the Middle-of-the-Road faction spurned cooperation and operated independently of the other parties. The Fusion platform of 1900 added an interesting item in its labor plank:

We denounce the practice of issuing injunctions in the cases of dispute between employers and employees, making criminal acts by organizations which are not criminal when performed by individuals, and demand legislation to restrain the evil.[13]

By the terms of the Norris-LaGuardia Act of 1932, federal courts were forbidden to issue injunctions against workers for striking, using union funds to aid the strike, furthering the strike by advertising, speaking, and picketing; holding mass meetings; and urging others to strike. The use of injunctions by federal courts in labor disputes did not again become an issue until passage of the Taft-Hartley Act of 1947.

It would, of course, be an exaggeration to give the Populists anything like full credit for the eventual enactment into law of many of their principal demands. As a party the Populists were not in existence when the laws were finally passed. The People's party may better be viewed as having given powerful impetus to the general reform movement which blossomed in the early part of the present century.

Among the principal results which may in part be attributed to the agitation and enthusiasm of the People's party are the following: the Australian ballot (used in all states except South Carolina); woman suffrage (19th Amendment to the Constitution); direct election of Senators (17th Amendment to the

[12] As quoted by Hicks, *op. cit.*, p. 444.   [13] Porter, *op. cit.*, p. 221.

PLATFORMS

Constitution); direct primaries (developed first on a statewide scale in South Carolina following the triumph of Ben Tillman in 1891); Presidential preference primaries (first used in Oregon in 1910); initiative and referendum (adopted by South Dakota in 1898); recall of elected officials (used chiefly in Western municipalities and in some Western states); flexibility in the currency supply (Federal Reserve Act of 1914); borrowing money against farm products (Warehouse Act of 1916); farm loan banks (created by Congress in 1916) and intermediate credit banks (1923); postal savings system (established in 1911); stabilization of farm prices and purchasing of surpluses (Federal Farm Board, 1929; Commodity Credit Corporation, 1933; Agricultural Adjustment Administration, 1933, 1938).

It should be noted that several important Populist demands have not become the law of the land. Notable among proposals which have apparently not found favor with the American people were those for government ownership of railroads and of the means of communication.

SOCIALIST MOVEMENT

The oldest socialist party still in existence is the Socialist Labor party, which was founded in 1876. Although the party has never had much of a following and has been split by various schisms, it has occasionally put forth pioneering demands of a practical nature. For example, the platform of 1892 called for:

Progressive income tax and tax on inheritances, the smaller incomes to be exempt.[14]

Official statistics concerning the condition of labor. Prohibition of the employment of children of school age and of the employment of female labor in occupations detrimental to health or morality. Abolition of the convict labor contract system.[15]

[14] Porter, *op. cit.*, p. 179.   [15] *Ibid.*

"The condition of labor" may now be ascertained from the charts, tables, and figures published by the Bureau of Labor Statistics of the Department of Labor. Limitations on the use of child and female labor have been established in all the states and the majority of such limitations have been upheld in challenges reaching the Supreme Court in recent years.

The Social Democratic or Socialist party of America presented its first Presidential ticket in 1900. The platform denounced capitalism as responsible for poverty, insecurity, and misery, and pressed for public ownership of the means of production.

In addition to this general program, which had not the slightest chance of electoral success, the Socialists endorsed a series of "immediate" demands which are of far greater significance in a study of American politics. These included a graduated income tax, direct election of Senators, prohibition of the use of injunctions in labor disputes, public works in time of depression, and state and national unemployment, accident, and old-age insurance.

Perhaps in retrospect the most significant among the Socialist demands of 1900 were the following:

The inauguration of a system of public works and improvement for the employment of the unemployed, the public credit to be utilized for the purpose.[16]

National insurance of working people against accidents, lack of employment, and want in old age.[17]

The adoption of the initiative and referendum, proportional representation, and the right of recall of representatives by the voters.[18]

It may be noted that the language of the second proposal is almost Rooseveltian, and that the Social Security Act of

[16] Porter, *op. cit.*, p. 241. The Populists made a similar proposal in 1896.
[17] *Ibid.*, p. 242. [18] *Ibid.*

August, 1935, is directed at an alleviation of two of the three evils mentioned.

An examination of Socialist party planks—especially for the period from 1904 to 1912—reveals a further resemblance between Socialist proposals and our current public law. The high point in Socialist electoral strength was reached in the Presidential election of 1920, when the vote for Eugene V. Debs amounted to more than 900,000 ballots. In 1924 the Socialists endorsed the candidacy of Senator LaFollette.

THE PROGRESSIVES OF 1924

The 1924 Presidential campaign of Senator Robert M. LaFollette took on added significance when the executive council of the American Federation of Labor came out in support of the Wisconsin Progressive. The threat was raised that all union labor might support him. LaFollette and his running mate, Senator Burton K. Wheeler, were forced to run on a variety of party labels, including Progressive, Independent, Independent-Progressive, and Socialist. The Communists ran their own candidate, while the Farmer-Labor party gave its support to LaFollette.

The Progressive candidate received close to five million votes, but he carried only Wisconsin with its 13 electoral votes. His platform is of considerable interest in the light of subsequent developments under the Roosevelt New Deal. Beginning with general remarks on the nature of democracy, the platform then included specific planks calling on the federal government to crush monopoly, to maintain freedom of speech and of press, and to take over the ownership of public utilities. The platform likewise demanded government ownership of railroads, the creation of a federal marketing corporation, "protection and aid of co-operative enterprises by national and state legislation," and the abolition of judicial review by the Supreme Court.

It was also proposed that there be direct nomination and election of the President, that the Constitution be amended to extend the initiative and referendum to the federal government, and that there be a popular referendum to decide whether the country should enter or refrain from entering any war in the future. In the perspective of today perhaps the most striking proposals were those dealing with the Supreme Court, the use of injunctions, and the practice of collective bargaining.

The court proposal read as follows:

Abolition of the tyranny and usurpation of the courts, including the practice of nullifying legislation in conflict with the political, social or economic theories of the judges. Abolition of injunctions in labor disputes and of the power to punish for contempt without trial by jury. Election of all federal judges without party designation for limited terms.[19]

The plank on collective bargaining was as follows:

Adequate laws to guarantee to farmers and industrial workers the right to organize and bargain collectively through representatives of their own choosing for the maintenance or improvement of their standards of life.[20]

The court proposal is reminiscent of the attempt by President Roosevelt in 1937 to "reform" the court. While Roosevelt lost the battle, he eventually won the campaign and was able to replace judges who had resigned with men he felt to be sympathetic to the New Deal program. The LaFollette plank on collective bargaining uses almost the identical phraseology of the Wagner Act of 1935. This phraseology was repeated in the Taft-Hartley Act of 1947.

RECENT STATE PARTIES

In recent years several state parties of a farmer or labor type have met with considerable success in securing the pas-

[19] MacKay, *op. cit.*, p. 271.   [20] *Ibid.*

PLATFORMS

sage of laws which they have sponsored. This has particularly been the case in those areas at one time dominated by the Non-Partisan League. Founded in 1915 in North Dakota by A. C. Townley, a former Socialist, the League attempted to promote cooperation between organized farmers and organized labor.

In Minnesota such cooperation reached sizable and effective proportions. Here the League in 1918 succeeded in nominating 61 of its 84 candidates for the lower house and 30 out of 43 for the senate. Labor organizations of the twin cities of Minneapolis and St. Paul gave wholehearted support to the League. Beginning in 1922 the League backed a new third party, known as the Farmer-Labor party, as the result of a working agreement with the Working People's Political League. From 1922 to 1938 the Farmer-Labor party enjoyed striking successes in Minnesota. Its period of greatest strength lasted from 1931 to 1939, although at no time did it have control of both houses of the state's legislature, which body was chosen at non-partisan elections. The victory of Harold Stassen in the 1938 gubernatorial race over both Farmer-Labor and Democratic opponents foreshadowed the end of one of the most interesting third party developments in American political history. In 1944 the party entered a coalition with the Democrats as the Democratic-Farmer-Labor party.

In North Dakota the Non-Partisan League captured control of the Republican party and has held control of that state for the greater period of time since 1915. The League embarked on a policy of agrarian "socialism," which included state-owned banks, flour mills, elevators, cooperative marketing, state crop insurance, minimum wage and workingmen's compensation laws, and a state-owned Home Building Association. The use of the state's taxing power to carry out these projects was, naturally, challenged in the courts. The validity of this type of state legislation was upheld, however, by the

United States Supreme Court in the case of *Green* v. *Frazier* (253 U.S. 233; 1920).

Many of the North Dakota laws, including the Home Building Association Act, the Mill and Elevator Association Act, the Bank of North Dakota Act, and the Industrial Commission Act, all passed in 1919, may have served as examples for the New Deal. In particular, the Federal Housing Administration and the Federal Deposit Insurance Corporation carried out on a national scale the type of activity previously tried on a local basis by North Dakota.

Farmer-labor parties have also been active in Wisconsin in recent years. When Philip LaFollette failed to obtain the Republican nomination for governor in 1932, he and his associates turned their attention to the formation of a new party. In 1934 the Wisconsin Progressive party was organized. The party sponsored legislation of a New Deal type in the state and cooperated with the Roosevelt Administration nationally through Wisconsin Progressives who had been elected to Congress.

In an effort to establish a national third party, Governor Philip LaFollette in 1938 announced the formation of the "National Progressives of America." The proposed program was to be similar in general tenor to the New Deal program in Washington. When the Farmer-Labor party of Minnesota and the American Labor party of New York refused to join the new alliance—primarily because they feared causing a split in the forces supporting President Roosevelt—Governor LaFollette abandoned his project.

In the fall election of 1938 Philip LaFollette was defeated and the Republican party took over the governorship. Senator Robert M. LaFollette, Jr., was reelected in 1940 as a Progressive. Subsequently party fortunes declined rapidly and in March, 1946, Senator LaFollette led the Progressives

PLATFORMS

back to the Republican fold. Despite this move he failed to win the Republican nomination for Senator in the spring primaries.

THE WALLACE MOVEMENT

The Wallace movement of 1948 continued in most respects the traditional methods of attack of farmer and labor parties. With the exception of their foreign policy planks, Wallace and his colleagues stressed the familiar farmer-labor demands: curbing of alleged monopolies, changes in those portions of the law dealing with labor relations, public ownership of various types of utilities, raising the income of the "common man," extension of social security and welfare legislation.

The continuity of ideas from the earliest parties down to the most recent is striking. Even the language of the parties of discontent has a familiar ring. The preamble to the 1892 platform of the People's party contains this denunciation:

The fruits of the toil of millions are boldly stolen to build up colossal fortunes for a few, unprecedented in the history of mankind; and the possessors of these, in turn despise the Republic and endanger liberty. From the same prolific womb of governmental injustice we breed the two great classes—tramps and millionaires.[21]

The Populist preamble is particularly harsh on the major parties:

They have agreed together to ignore, in the coming campaign, every issue but one. They propose to drown the outcries of a plundered people with the uproar of a sham battle over the tariff, so that capitalists, corporations, national banks, rings, trusts, watered stock, the demonitization of silver and the oppressions of the usurers may all be lost sight of. They propose to sacrifice our homes, lives, and children on the altar of mammon; to destroy the multitude in order to secure corruption funds from the millionaires.[22]

[21] Porter, *op. cit.*, p. 166.   [22] *Ibid.*, p. 167.

A quieter but similar tone can be found in the preamble to the LaFollette platform of 1924: "The great issue before the American people today is the control of government and industry by private monopoly." [23]

The speech of Henry A. Wallace at Chicago on December 29, 1947, in which he announced that he would be an independent candidate for the Presidency, contains much the same thought as the utterances of Weaver, Debs, and LaFollette. On that occasion Wallace told a nation-wide radio audience:

We have assembled a Gideon's army—small in number, powerful in conviction, ready in action. We have said with Gideon, "Let those who are fearful and trembling depart." For every fearful one who leaves there will be a thousand to take his place. A just cause is worth a hundred armies. . . . By God's grace, the people's peace will usher in the century of the common man.[24]

The preamble of the platform of the 1948 Progressive party started out with these words:

Three years after the end of the Second World War, the drums are beating for a third. Civil liberties are being destroyed. Millions cry out for relief from unbearably high prices. The American way of life is in danger.

The root cause of this crisis is big business control of our economy and government.

With toil and enterprise the American people have created from their rich resources the world's greatest productive machine. This machine no longer belongs to the people.

Never before have so few owned so much at the expense of so many.[25]

The Progressive platform itself included planks calling for cooperation with the Soviet Union, disarmament, various federal civil rights enactments, government ownership of the largest banks, the railroads, the merchant marine, and public

[23] Porter, *op. cit.*, p. 516.　[24] Associated Press dispatch, December 30, 1947.
[25] Associated Press dispatch, July 24, 1948.

utilities, a federal emergency housing program, repeal of the Taft-Hartley Act, extension of social security coverage, federal aid to education, and tax revisions to aid small income groups.[26] Criticism of the platform by persons opposed to the Progressive party was concentrated on those planks dealing with foreign policy, specifically, with the Soviet Union and with the United Nations. It was frequently charged that these planks were both pro-Soviet and isolationist.

The Progressives made every effort to identify themselves with Jefferson, Jackson, Altgeld, LaFollette, Norris, and Franklin Roosevelt. While such identification was by no means generally accepted by the public, it is significant that such an effort to establish a kind of spiritual continuity was made.

CONCLUSIONS

Four significant conclusions may be drawn from this analysis. The first of these is that farmer-labor parties have, historically, served as a tremendous impetus toward reform. It would be erroneous to assume that all impetus toward reform has come from such parties. Many of the ideas subsequently advocated by farmer and labor parties were initially popularized by non-partisan pressure groups, which were prepared to operate through any party which might be in power. Civil service reform has been mentioned as one case in point; another example was furnished by the activity of the National Labor Union in promoting the cause of Greenbackism. For nearly every reform proposed by the Populists the genesis of the idea can be traced to groups which existed before the Populist party was founded.

Yet the comparison between many of the planks found in the platforms of farmer and labor parties and the present public law of the United States shows a similarity which can-

[26] Full text of platform may be found in New York *Times* of July 25, 1948.

not be ignored. The general pattern is this: Discontent occurs over a specific issue. A farmer or labor party agitates on this issue. In time the agitation reaches such formidable proportions that a major party can no longer look past the issue but must face it. Thus, at the very time of its greatest success—acceptance of its main issue by a major party—a farmer or labor party tends to go out of existence.

A second conclusion is that the degree of continuity in the ideology of many of the farmer and labor parties is impressive. Monopoly and big business have generally been the enemies; government regulation, labor reform, and social legislation have generally been favored. In many cases even the language is similar. It is interesting to observe that both the Populists and the followers of Henry A. Wallace were prone to use Biblical quotations.

A third conclusion is that the existence of farmer and labor parties, no matter under what name, is too consistent to be dismissed as an occasional rash on the body politic. Politicians of the major parties are inclined toward such an analysis, but their judgments cannot always be accepted as being disinterested. Underneath the surface of American political life there has apparently always existed a layer of discontent. This discontent has taken varying forms but has usually been present. When the discontent has assumed dangerous proportions— that is, when it has become a threat to the major parties—it has been recognized and one or both of the major parties have committed themselves to remedial action. Discontent generated over the specific issue has then gradually disappeared, to be replaced in time by another issue of equal intensity, which eventually challenges the attention of the major parties. This has been the recurrent pattern of American politics.

One final conclusion remains to be stated. It can perhaps best be expressed in the form of a question: When viewed historically do the platforms of farmer and labor parties of the past appear to modern students of the political process to have been "radical"? In general the answer is "no." It must quickly be conceded, however, that the platforms of certain of the fringe or extremist parties—such as the Single-Taxers, the Communists, the Trotskyites—have in many ways been outside the American political tradition. Yet the great parties of the American farmer and labor movements—the Grangers, the Populists, the Socialists of the early part of the century, the Wisconsin and Minnesota and New York parties of the more recent era—have by and large operated within the American tradition. The adoption by the Congress and state legislatures of so many of their planks attests to this fact.

3

THE RECORD AT THE POLLS

PRESIDENTIAL ELECTIONS

IF THE TITLE,"most exciting show on earth," could be taken away from our largest circus, it might well be applied to the quadrennial Presidential elections. These elections attract the maximum in interest, oratory, money, and, of course, votes. Since more Americans vote in Presidential than in other elections, it is understandable that farmer and labor parties usually enter these contests in order to test their strength on a national basis.

There have been times in our history when Presidential elections showed a relatively small difference in the vote cast for the two major candidates. In many of the elections before 1900 it was mathematically possible for farmer and labor parties to hold a balance of power in popular votes. For example, in the election of 1880 the Republican majority over the Democratic candidate amounted to 9,464 votes. At the same election the Greenback party, under General Weaver, received 308,578 votes. In terms purely of popular votes, the Greenbackers held the potential balance of power. In 1884, 1888, and 1892, the vote for the candidate of the chief farmer or labor party amounted to more than the difference in votes given the candidates of the two major parties.

This situation has changed drastically in the twentieth century. The only year when a farmer or labor party approached holding the balance of power was 1916. In that year the Democratic lead of 594,000 was only 4,000 more than the Socialist party vote. But in 1936 the difference between the vote cast for the major parties was nearly eleven million; in 1940 it was nearly five million; and even in 1944, a relatively close year for the present century, it amounted to more than three and a half million votes. These differences are so great that, even in mathematical terms, farmer and labor parties have been unable to attract a sufficient vote to hold the balance of power.

The total votes for farmer and labor parties in Presidential elections are shown in Table 1.[1] A significant feature illustrated by the table is that in every one of the past twenty Presidential elections one or more parties of the farmer or labor type have competed.

Also of significance is the longevity of some of the parties. The Socialist Labor party has competed in every election since 1892. Likewise illustrated by the same table is the tremendous elasticity of the vote for farmer and labor parties. This elasticity is particularly demonstrated by a glance at the vote for the Socialist party, which in 1920 polled more than 900,000 votes for its slates of Presidential electors. The all-time high was registered, however, by the LaFollette Progressives of 1924, with nearly five million votes. The other peak in farmer-labor voting was reached in 1892, when the Populists ran General Weaver for President.

The sustained level of voting for farmer and labor parties in Presidential elections since 1928 is accounted for as follows: In 1936 the Union party, running on a farmer-labor platform (although its leadership was considered by many to be protofascist) amassed a total of 892,000 votes; in 1936, 1940, and 1944 the American Labor party of New York State received 274,000, 417,000, and 496,000 votes respectively; in 1944 the Liberal party of New York State received 329,000 votes. Both

[1] Table 1 has been compiled from the following sources: *McPherson's Handbook of Politics, Tribune Almanac, American Almanac,* from 1872 to 1892; E. E. Robinson, *The Presidential Vote, 1896–1932* (Palo Alto: Stanford University Press, 1934), from 1896 to 1932; E. E. Robinson, *They Voted for Roosevelt* (Palo Alto: Stanford University Press, 1947), from 1936 to 1944. It is extremely difficult to get accurate election statistics for the period prior to 1896.

TABLE 1

VOTE FOR FARMER AND LABOR PARTIES IN PRESIDENTIAL ELECTIONS, 1872–1948
(In Thousands of Votes)

Year	Greenback	Union Labor	Populists	Socialist Labor	Socialist	Farmer-Labor	LaFollette	Communist	Liberty	Union	American Labor	Liberal	Wallace†	Total of All Farmer and Labor Parties*
1872														29
1876	82													82
1880	309													309
1884	175													175
1888		148												151
1892			1,041	21										1,062
1896			124	35										159
1900			50	33	93									176
1904			113	33	403									549
1908			29	14	420									463
1912				27	900									927
1916				15	590									605
1920				30	911	265								1,206
1924				38			4,831	29						4,898
1928				22	268	7		48						345
1932				34	873	7		101	53					1,070
1936				13	188			79		892	275			1,447
1940				15	117			49			417			598
1944				45	79						496	329		950
1948				28	132						[510]	223	1,157	1,553

* Total for 1888 includes United Labor, 2,808; for 1932 includes Jobless, 740; for 1948 includes Socialist Workers, 13,000. Figures for 1948 are preliminary.

† Wallace total includes ALP vote in New York State.

THE RECORD AT THE POLLS 35

the Labor and Liberal parties supported Franklin Roosevelt in these elections. In 1948 the Wallace party received 1,157,000 votes, while the American Labor party and the Liberal party received 510,000 and 223,000, respectively.

Chart I expresses the combined votes of farmer and labor parties as percentages of all votes cast in each Presidential election since 1872.[2] This chart illustrates a point that has already been made in another connection—the peak protest

CHART I

VOTE FOR FARMER AND LABOR PARTIES IN PRESIDENTIAL ELECTIONS, 1872–1948

(Combined votes for head of tickets of all farmer and labor parties expressed as per cent of total vote.)

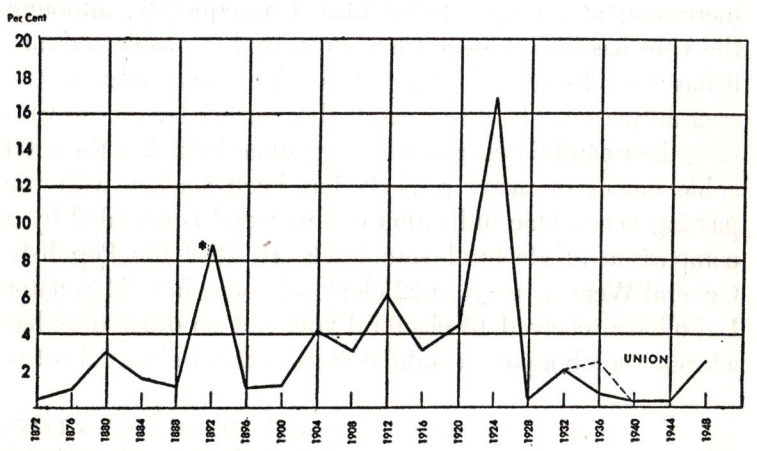

* In 1892 Populists fused with one or the other major party in some states. The combined vote is here included.

[2] Chart I has been compiled from the same sources as Table 1, with the total vote of all farmer and labor parties divided by the total number of votes cast in each Presidential election. This shows the share of the total vote received by farmer and labor parties. The Union party of 1936 is shown with dotted lines in order to reveal the figures with and without the inclusion of the vote of that party.

years were 1892 and 1924. Discontent also reached a high level in 1912. From 1872 to 1924 the general trend in protest voting was upward. During the Rooseveltian elections of 1936, 1940, and 1944, electoral discontent was relatively stabilized, if the percentage of votes cast for farmer and labor candidates is used as an indication. The 1912 percentage reflects largely the tremendous Socialist vote of that year. The near-even percentages during the 1936, 1940, and 1944 elections are to be accounted for by the vote for the Union party, the American Labor party, and the Liberal party.

From the foregoing discussion some assessment of the showing of farmer and labor parties in Presidential elections may be made. The absolute vote won by them has been, in general, increasing since 1872. But the size of the electorate has been increasing at an even faster rate. Consequently, although the vote has been bigger when expressed in absolute terms, it has not substantially changed in relative terms during the over-all period. The farmer and labor parties have been running their candidates in more states since 1872, but the total achievement in many respects has been smaller with the passing years. One indication of this trend is revealed by a comparison of electoral vote totals. In 1892 the Populists' General Weaver received 22 electoral votes; in 1924 Senator LaFollette received 13 electoral votes; at no other time has a farmer or labor party candidate received any electoral votes.

A FOOTNOTE ON CLOSE PRESIDENTIAL ELECTIONS

Attention has been called to the fact that the period from 1876 to 1892 is characterized by close elections and an even distribution of the popular vote between the major parties. But since elections are decided by electoral, not by popular, votes, it is difficult to show instances beyond doubt where a minor party did succeed in changing the outcome of an election. One possible case appears to be the election of 1884. Benjamin Butler's Greenback vote of 16,994 in a mathematical sense may have swung New York to Grover Cleveland and the Democrats. The final electoral vote was 219 to 182, in favor of Cleveland over Blaine. Cleveland carried New York by a plurality of 1,149 and the 36 electoral votes of New York proved decisive in the election. It should be pointed out, however, that if Blaine had not neglected to rebuke Burchard's description of the Democratic party as the party of "rum, romanism, and rebellion," he might have carried New York State. In addition, the Prohibition vote—which was larger than that cast for Butler's Greenback party—would also have carried New York had the Prohibitionists supported Blaine against Cleveland instead of running their own candidate. Thus two parties potentially held the balance of power in deciding which major party should carry New York, and hence the national election.[3]

STATE ELECTIONS

In addition to showing the percentages of all votes cast that were given to farmer and labor Presidential candidates, Chart II also summarizes the vote for the head of the ticket of such parties in non-Presidential years.[4] Since there are usually fewer votes cast in non-Presidential elections than in Presidential contests, a percentage comparison is the only equitable method of contrasting the statistics. In compiling

[3] The victory of Cleveland in New York State in 1884 is often attributed to the vote polled by the Greenback party. This, it is said, took the state away from Blaine. For example, see Cortez A. M. Ewing, *Presidential Elections, from Abraham Lincoln to Franklin D. Roosevelt* (Norman: University of Oklahoma Press, 1940), p. 134. Such an interpretation has the virtue of simplicity but is of questionable accuracy.
[4] Only figures for even-numbered years are presented in this chart, because it was found that in most states, even in the early years, few significant elections occurred in odd-numbered years. Voter participation was always light in these off-year elections, a trend which became even more apparent after 1926.

Chart II the technique used was to take the highest vote in each state (usually that for governor), calculate the total vote of such candidates for all states combined, and, in each election year, express this vote as a percentage of the total vote for all parties.

CHART II

VOTE FOR FARMER AND LABOR PARTIES IN PRESIDENTIAL AND STATE-WIDE ELECTIONS, 1872–1948

(Combined votes for head of tickets of all farmer and labor parties expressed as per cent of total vote.)

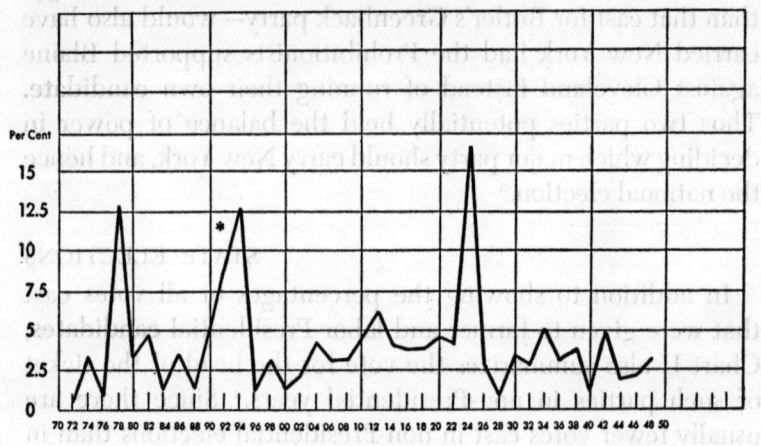

* In 1892 and 1894 the Populists fused with one or the other major party in some states. From 1896 on, no Fusion votes are included. Total for 1896 includes Bryan's vote on the Populist ticket only.

Such a technique brings out factors which are not evident from a study of Presidential elections alone. The peak of the Greenback movement was reached in the general elections of 1878, when that party received 12.7 per cent of all votes cast in elections for state-wide office. The height of the Populist movement was reached in 1894, when Populist candidates were accorded a vote equal to 12.5 per cent of all the vote.

THE RECORD AT THE POLLS

(The voting statistics for 1894, as well as for 1892, include votes for Fusion candidates.) It is interesting to observe that in 1934, 1938, and 1942—all non-Presidential years—the farmer and labor candidates running for state-wide office did better than farmer and labor Presidential candidates running in 1932, 1936, and 1940. The explanation for this phenomenon lies in the fact that such parties as the Wisconsin Progressives, Minnesota Farmer-Labor party, and certain smaller groups supported Franklin D. Roosevelt in Presidential elections but ran their own candidates in state elections. It would thus appear that the principal victories of many farmer and labor parties should be looked for on the state, instead of on the national, level.

The record of the Minnesota and Wisconsin parties bears out this assumption. The Minnesota Farmer-Labor party, which came into being largely as the result of Non-Partisan League activities, contested the governorship in every election beginning with that of 1922. From then until 1944 the Farmer-Labor party was generally the leading opponent of the Republican party, with the Democrats running a decidedly inferior third. The Farmer-Labor party elected its gubernatorial candidate in 1930, 1932, 1934, and 1936. Since the problem of the Democratic party was to stay alive, it frequently supported the Farmer-Labor party candidates.

The Farmer-Labor party won Senatorial elections in 1922, 1923 (special election), 1928, 1934, and 1936. It won 24 seats in the House of Representatives during this same period. The greatest numbers of Representatives were in the 73d Congress (five), elected in 1932, and in the 75th Congress (five), elected in 1936.

The 1938 gubernatorial election resulted in an overwhelming victory for the Republican candidate, Harold Stassen, over both Farmer-Labor and Democratic candidates. Two years later Senator Shipstead, until then perennially successful as a Farmer-Labor candidate, turned Republican and was reelected to the Senate as a member of that party. Despite the creation of a Democratic-Farmer-Labor coalition party in 1944, the Republicans triumphed in the gubernatorial contest of that year and also those of 1946 and 1948. But in 1948 the coalition party elected Hubert H. Humphrey, an official of the liberal Americans for Democratic Action, to the United States Senate. The same party won four of the nine Minnesota seats in the House of Representatives.

The role played by the Progressive party of Wisconsin, while briefer, has been equally spectacular. The Progressives won the gubernatorial elections in 1934 and in 1936. They won Senatorial elections in 1934 and in 1940. In all four cases a LaFollette was the successful candidate. By 1946 the LaFollette name had apparently lost some of its magic. In the spring of that year Senator Robert M. LaFollette, Jr., led the Progressives back into the Republican party. Despite this move he failed to receive the Republican nomination for Senator and temporarily bowed out of politics.

During the period of their ascendancy the Progressives of Wisconsin won 22 seats in the House of Representatives. In the 1934 and 1936 elections they swept the state, electing seven representatives as the result of each election.

As illustrations of straight power politics—trying to win elections on the winner-takes-all principle—the American Labor party of New York State, and the Liberal party of the same state, are not comparable to the Wisconsin and Minnesota parties of the 1930's. The balance-of-power role of the American Labor and Liberal parties is discussed elsewhere. Yet, the New York parties have amassed sizable votes, and the American Labor party has sent two of its candidates to Congress.

In 1936 the American Labor party, which joined the Democrats in supporting Herbert Lehman, received more than

THE RECORD AT THE POLLS

260,000 votes. In 1938, in its second gubernatorial contest, the ALP rolled up 419,000 votes. In 1946 it contributed 425,000 votes to the Democratic-ALP-Liberal candidate for governor. Its record in New York City contests has been equally impressive. In 1937 it gave LaGuardia, as candidate for mayor, 482,000 votes; in 1945 it contributed 259,000 votes to the Democratic-ALP candidate, William O'Dwyer. The Liberal party, which was formed in 1944, rolled up 122,000 votes for the Republican-Liberal mayoralty candidate in 1945. The same party gave 180,000 votes to Mead, the Democratic-ALP-Liberal candidate for governor, in the 1946 elections.

From the 76th through the 81st Congress the ALP has had representation in the House of Representatives. The total stood at one Representative per Congress until the 80th Congress, when victory in a special election in the 24th Congressional district of New York on February 17, 1948, raised the total to two for the 80th Congress. As a result of the 1948 elections the ALP representation in the House declined from two to one. To date the Liberal party has elected only one Congressman—Franklin D. Roosevelt, Jr., from the 20th Congressional district, at a special election held on May 17, 1949.

Any statistical comparison between the earlier Granger and Populist movements and more recent parties in terms of gubernatorial elections is difficult and perhaps misleading. This is because the earlier agrarian parties frequently resorted to fusion tactics in state-wide elections. It may be stated with certainty, however, that the more recent parties have not approached the great agrarian parties of the last century in ability to elect state-wide officers. It should likewise be noted that the Granger and Populist movements covered a far wider area than the more recent parties, which have tended to be one-state organizations.

These statements are not intended to belittle the achievements of some of the state-wide parties of modern times. During its heyday the Wisconsin Progressive party was as successful, electorally, as any Granger or Populist group. Yet a comparison in terms of Senators and Congressmen elected shows the relative weakness of the recent parties. In 1937 some 13 Representatives and three Senators were elected to Congress by farmer and labor parties. This is a slight total indeed when compared to the figures for the 55th Congress. After the 1897 elections the totals for Senators and Congressmen of farmer and labor parties (including Fusion candidates) stood at eight and 28 respectively. Yet for sheer consistency over a period of time, the prize must go to the Minnesota Farmer-Labor party. It had continuous representation in Congress from 1923 to 1944, a period of 22 years. This far surpassed the record of the Populists.

Farmer and labor parties of the twentieth century have experienced far less success than the agrarian parties of the last century in their attempts to control state legislatures. One of the reasons for this is that protest parties today tend to be urban and labor, instead of rural and farmer as formerly. From the point of view of the urban protest party the problem of controlling a state legislature, even in many heavily industrialized states, may be insoluble. This may be the case even though it is mathematically possible for such a party to win control of the governorship and other state-wide offices.

The explanation of this paradox lies in the underrepresentation of urban areas in the legislatures of many of the states.[5] A few examples illustrate the difficulties encountered. Cook County, Illinois, had in 1940 more than one-half the population of the state. Had apportionment been on the basis of population, the county would have had control of both houses

[5] The interested reader will find an excellent discussion of urban underrepresentation in state legislatures in Charles M. Kneier, *City Government in the United States* (New York: Harper, 1947), pp. 109–131.

of the legislature. But since no reapportionment has been made since 1901, the representation of Cook County has remained at 19 of 51 senators and 57 of the 153 representatives.

The situation is similar in New York State. New York City, with 54.5 per cent of the total population of the state and 53.4 per cent of the citizen population (the basis on which apportionment is made), has only 44.6 per cent of the membership in the state senate and state assembly. Additional examples could be given from California, New Jersey, Michigan, Missouri, and other states to demonstrate that as a general rule urban areas are deliberately underrepresented in state legislatures.

There are many reasons for this. Some are purely political, others are social, racial, and religious. The effects of underrepresentation are widespread, but it should be noted that purely urban protest parties operate under such a system at a tremendous disadvantage. A real farmer-labor party, drawing support from both farmers and from urban workers, rarely exists. In practice most recent protest parties have found the bulk of their strength in urban areas. Thus, the Granger and Populist parties had an advantage over their modern-day counterparts in their greater ability—actual as well as potential—to win control of state legislatures.

CONGRESSIONAL ELECTIONS

The record of farmer and labor parties since 1872 tends in some respects to be more impressive in Congressional than in Presidential elections. This is caused by the fact that many of the parties have been regional in character. Thus, some parties have on occasion been able to carry Congressional districts with ease, while at the same time they encountered insuperable obstacles in presenting a Presidential candidate. The very frequency of elections for the House of Representatives also tends to make such elections a more satisfactory mirror of public sentiment than the quadrennial Presidential contests. Because only one-third of the Senate is elected at one time, Senatorial elections are less satisfactory as means of measurement than House elections. In addition, before 1913 (when the 17th Amendment went into effect) Senators were elected by vote of the legislatures of the states. (In some Western, Midwestern, and Southern states the legislatures had adopted long before 1913 the practice of permitting the voters to express a choice for Senators. The legislatures then proceeded to ratify the popular mandate.)

The general story of Congressional elections is shown in Chart III.[6] In contrast to other charts where Fusion votes are not included after the merger of the Democrats and the Populists on a national scale in 1896, the chart of Congressional elections includes Fusion candidates through the 1890's and the early twentieth century. The reason for this inclusion is that it is impracticable to separate Democratic-Populist and Populist candidates in Congressional elections. In every case, however, only candidates who successfully ran on farmer or labor party tickets are included. (By reference to Appendix B the reader may identify for himself the parties whose candidates have been included in the compilation.)

Three principal features of American politics are illustrated

[6] Both Chart III and Appendix B—representatives of farmer and labor parties elected to Congress—have been compiled from the *Congressional Directories* of the various Congresses, with such additional information as was contained in the standard almanacs of each period. The criterion in every case was whether the successful candidate ran under the auspices of a farmer or labor party and carried the label of such a party. In contrast to the other charts and tables referring to voting statistics, Fusion candidates running for office after the merger of the Democrats and the Populists in 1896 have been included. The Congressional districts from which the Congressmen were elected have not been given because periodic reapportionment and redistricting would make such information of questionable value. Since party names are given, the reader may analyze the party designations for himself, together with the geographical strength of each party. To find the limits of any Congressional district at any given time, the simplest method is to consult the *Congressional Directory* of the appropriate Congress.

by an analysis of Congressional elections: the persistency of certain farmer or labor parties; the regional characteristics of many of the parties; and the better showing made by the relatively non-ideological parties than by parties possessing a Marxist philosophy.

CHART III

NUMBER OF CANDIDATES OF FARMER AND LABOR PARTIES ELECTED TO THE UNITED STATES HOUSE OF REPRESENTATIVES, 1870–1948

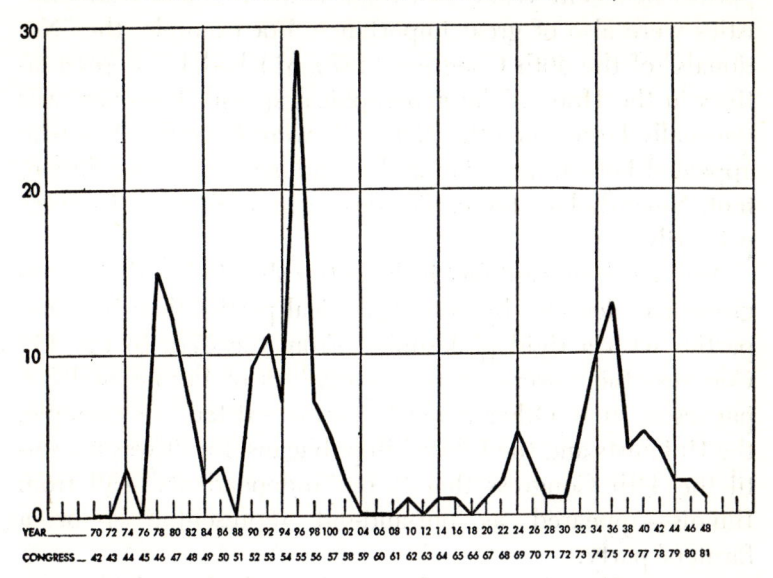

The same persistency which was shown in the analysis of Presidential elections is again present in Congressional elections. Despite some popular belief to the contrary, in certain areas farmer and labor parties have enjoyed a fairly continuous existence. The Wisconsin Progressives held seats in every Congress from the 76th through the 79th; the Minnesota Farmer-Labor party held seats in every Congress from the 68th through the 78th; the Socialists held seats from the 67th through the 70th Congress, and also were represented in the 62d, 64th, and 65th Congresses. With representation from the 76th through the 81st Congress, it appears that the American Labor party of New York State is following a well-beaten track.

Merely to mention the names of the successful parties is to imply a show of regional strength. Wisconsin, Minnesota, and New York have been seats of strength for farmer and labor parties in recent years. In earlier years, regional characteristics were also of great importance. For example, the "Nationals" of the 46th Congress (1879–81) had 15 representatives in the House. The geographical spread, however, was unusually large since the Nationals were Greenbackers who appealed both to agrarian and to industrial areas of discontent. Nevertheless, the earlier farmer and labor parties were primarily regional.

Another significant fact is the better showing made in Congressional elections by non-ideological parties than by those parties with a rigid philosophy. From the 37th to the 43d Congress there were no representatives of farmer or labor parties as such. Other minority parties existed, for example, the Unionists and the Liberal Republicans. But it was not until the 44th Congress that three "Independents"—all from Illinois—appeared as out-and-out representatives of a farmers' party.

During the entire period from the 44th through the 81st Congress a total of 49 Senators and 175 Representatives served as members of farmer or labor parties. But the only party with a relatively fixed ideology (by American standards) to achieve any successes at all was the Socialist party. These successes have not been great in comparison with the record of other farmer and labor groups. The Socialists have elected eight Representatives and no Senators to this point in our history. Of the eight House elections won by the Socialists, four were won in Wisconsin and four in New York. The

Socialist record stands in marked contrast to that of the Grangers, Populists, Farmer-Laborites of Minnesota, and Progressives of Wisconsin. Marxist parties to the left of the Socialists have enjoyed no successes in Congressional elections.

In examining Chart III it is well to keep in mind that the size of the House, as well as of the Senate, has changed from time to time. Thus, the 15 farmer-labor members of the House in the 46th Congress (elected in 1878) represented 5.1 per cent of the House; the 28 farmer-labor representatives of the 55th Congress (elected in 1896) represented 7.8 per cent of the total membership; the 13 farmer-labor Representatives in the 75th Congress (elected in 1936) represented only 2.9 per cent of the House membership.

One final implication of the voting for farmer and labor parties in Congressional elections deserves comment. Analysis of the Congressional districts sending farmer or labor representatives to Congress since the 44th Congress suggests that the strength of such parties has been shifting to the industrial areas. This suggestion is confirmed later in this study by an examination of regional patterns in farmer-labor voting.

CONCLUSIONS

From the data presented in this chapter some definite conclusions emerge. A striking characteristic of farmer and labor parties is their lack of permanence under any given name. Except for the Socialist Labor and the Socialist parties, no farmer-labor party has survived more than five Presidential elections. Another characteristic of these parties is that they have little strength at the polls. Unlike the major parties, which rarely poll less than 30 per cent of the total vote cast, farmer-labor parties show great fluctuations in strength. The record reveals a series of spurts followed by collapses.

The relatively short life of individual parties and the great variations in the number of votes attracted should not, however, mislead one into predicting that the day of third parties is over. Professional politicans and many political commentators are prone to opine after certain elections that the day of farmer-labor parties has gone forever. Viewed in the perspective of history, such a prediction is dangerous. The years in which farmer or labor parties have done well at the polls have followed periods of little success. Thus Streeter, running under the label of the Union Labor party, was able to capture only one per cent of the total vote cast in 1888. Yet, the Populists' General Weaver rolled up 8 per cent of the total vote only four years later.

The amazing showing of the Populists in the 1892 Presidential election and in the state elections of 1894 was followed after 1896 by the almost complete disappearance of the party. After 1896 the Populists never again polled so much as one per cent of the popular vote for President. The same phenomenon is to be observed in the case of the 1924 Progressives. After obtaining nearly five million votes in the Presidential election, the party fell apart with the death of Senator LaFollette the next year. In 1928 the combined votes of the remaining farmer-labor parties amounted to less than one per cent of all votes cast.

On the basis of recent Presidential elections, statements have again become fashionable to the effect that farmer-labor parties will be forever eclipsed. Such statements miss the principal point. Historically, the vote for farmer-labor parties has not been dependent upon a mere perfecting of political machinery and organization; it has been dependent on conditions making for economic and social discontent.

Keeping in mind the historical perspective since 1872, there is evidence that despite violent ups and downs, farmer-labor parties are capturing an increasing share of the popular vote over a period of time. Yet the patterns of concentration

THE RECORD AT THE POLLS

and distribution of the vote indicate that these parties are less effective today than they were in the nineteenth century.

There have been only four occasions when a farmer or labor party polled a significant portion of the total national vote. These occasions were the state elections of 1878, when the Greenbackers received 12.7 per cent of the vote; the Presidential election of 1892, when the Populists received 8.7 per cent of the total vote; the state elections of 1894, when the Populists obtained 12.5 per cent of the vote; and the Presidential election of 1924, when the Progressives were given 16 per cent of the total vote.

Some of these elections contain lessons for the major parties. In 1924 the Democratic party polled only 28.8 per cent of the votes cast in the Presidential election. In that year it ran an extremely conservative candidate. Thus a good part of the protest vote went to Senator LaFollette, while most of the conservative vote went to the Republican candidate. To win a Presidential election the Democratic party must absorb a considerable portion of the protest vote. In 1932 the Democrats successfully did this, and the protest vote as expressed in voting for farmer and labor parties was small. In 1948 a good share of the early supporters of Wallace eventually voted for Truman.

Finally, it must be observed that only those farmer and labor parties with moderate ideologies have done well at the polls. The record clearly demonstrates this. The "radical" or extremist parties have not appealed to any sizable percentage of the American electorate. Unless the general social relationships of the country should undergo profound changes, it appears unlikely that extremist varieties of farmer-labor parties will prosper at the polls. Success—relative though it may have been—has belonged to the moderates.

4

GEOGRAPHIC PATTERNS OF PROTEST VOTING IN PRESIDENTIAL ELECTIONS

SINCE 1872, when the first farmer-labor party competed in a Presidential election, tremendous population movements and changes in the national economy have taken place. One of the most important changes, from the point of view of farmer-labor party voting, has been the gradual decline of the open-country and village population as compared to city and town dwellers. In 1880, during the heyday of the Greenback party, some three fourths of our population lived in the country. By the 1920's, when LaFollette tried to combine farmers and workers on behalf of concerted political action, more than half the population was urban. "Urban" was defined to mean towns and cities of 2,500 or more persons. The trend toward urbanization has continued. By 1940 about half of the people of the United States lived in urban places of more than 10,000 persons, and a decided concentration had occurred in cities of more than 100,000 persons and in metropolitan districts.

The decline in rural population has reflected both the relative diminution of agriculture as a factor in the total economy and the increased importance of industry. By the start of the present century the shift was clearly apparent. For the first time the number of industrial wage earners exceeded the number of farmers and farm hands.

With this basic population shift in mind, what can be said of the trends in voting for farmer and labor parties since the election of 1872?

One measure of absolute strength is the actual number of votes cast for farmer and labor parties. An examination of

GEOGRAPHIC PATTERNS

the distribution of these votes state by state points to the following general conclusions:

Until 1900 the centers of discontent shifted from one geographic region to another. In most of the Presidential elections to 1900 the largest number of votes was cast in the states bordering the Great Lakes, the states of the great wheat belt in the West North Central area, and the states of the South. In the last category are included, although with some fluctuations in voting, both cotton and range states. The only notable exception to the general concentration of votes in farm centers occurred in 1884. In that year General Weaver received a third of his votes from Massachusetts, New York, and Pennsylvania.

A brief survey election by election illustrates the extent of shifting in the protest area of the West and South prior to 1900.

In the Presidential elections of 1872, 1876, and 1880, farmer and labor parties were strongest in actual voting strength in the corn, wheat, and hog belts. In 1872 the seven states of Iowa, Missouri, Illinois, Kentucky, Indiana, Ohio, and Michigan accounted for 52 per cent of the vote given the Labor Reform party. Four years later 64 per cent of the vote cast for the Greenbackers came from Illinois, Michigan, Indiana, Iowa, and Kansas.

By 1884 a significant amount of electoral discontent with the major parties was manifested in the Eastern states. Of the total vote amassed by General Weaver on the Greenback ticket, 36 per cent came from Michigan, Illinois, and Indiana, while another 35 per cent came from Massachusetts, New York, and Pennsylvania.

Agrarian discontent ran high in 1888. The contender for the farmer-labor votes—the Union Labor party—polled 58 per cent of its total vote in the three states of Kansas, Texas, and Missouri. Four years later, with the farm depression worse, the Populists entered their first Presidential race. Their best states were all in the Midwest, Southeast, and the Southwest. Colorado, Nebraska, Kansas, and Missouri accounted for some 33 per cent of the Populist vote, while Texas, Alabama, and Georgia collectively accounted for 26 per cent.

Such was the situation before 1900. Since then the voting for farmer and labor parties has clearly reflected the national population shift from rural to urban areas. At least half of the votes have come from the heavily populated states of the Middle Atlantic, East North Central, and Pacific areas. Simultaneously the great centers of Populist strength—Kansas, Missouri, Nebraska, and the South—have dropped out of the picture.

First to feel the effects of a shift from the old West and the South to the Northwest and the Northeast was the Socialist party. In the South it managed for a time to pick up the remnants of the Populist party, but gradually its electoral support in that area dissolved and its strength shifted to the Great Lakes and Middle Atlantic areas. This distribution pattern in terms of absolute votes has also characterized the voting for other farmer and labor parties since 1900.

The Farmer-Labor party of 1920, running Christensen as its Presidential candidate, did surprisingly well in the Middle Atlantic and East North Central states. While a good share of Christensen's strength came from Washington and South Dakota, yet the combined votes from Illinois, Michigan, Indiana, New York, and Pennsylvania accounted for 42 per cent of the candidate's total vote. In the election of 1924 La-Follette got the largest share—73 per cent—of his votes from New York, Pennsylvania, Massachusetts, Ohio, Michigan, Indiana, Illinois, Wisconsin, Minnesota, Iowa, California, and Washington.

The same general pattern is observable in the distribution of the vote for Lemke's catch-all Union party in 1936. Illinois,

GEOGRAPHIC PATTERNS 53

Michigan, Indiana, and Pennsylvania gave Lemke 47 per cent of his entire vote. Had the Union candidate been on the ballot in New York, he undoubtedly would have added substantially to his over-all total.

Finally, it should be pointed out that the great state-wide farmer-labor parties of the Roosevelt era were located in the Great Lakes area and in New York State. These parties were the Farmer-Labor party of Minnesota, the Progressive party of Wisconsin, and the American Labor and Liberal parties of New York. In New York State the protest vote has been concentrated in New York City. In 1948 some 87 per cent of the Wallace voters lived in the states of New York, California, Pennsylvania, Michigan, New Jersey, Massachusetts, Ohio, Washington, Minnesota, and Wisconsin. About half of Wallace's support came from New York State alone.

The distribution of actual votes won in Presidential elections is not the only measure of the strength of these parties. This method tends to weight heavily those states having the largest populations. That is, the sheer weight of population is a very important factor in explaining the concentration of absolute votes for farmer and labor parties in certain areas. An alternative measure of strength is the per capita vote in each state. This method is intended to estimate the intensity of voting for farmer and labor parties by determining what per cent of the entire vote was cast in each state for such parties. The unit of comparative measurement is the state, since the states, no matter how artificial their boundaries may be, constitute the voting units in Presidential elections.

An analysis of the voting history of each state since 1872 reveals patterns in voting for farmer and labor parties. In grouping the states, emphasis was placed upon the elections of 1892 and 1924, since these were the only elections in which farmer-labor parties made a significant showing of strength. Other elections, of course, enter into the analysis, including the election of 1948. When the elections of 1892 and 1924 are used as the principal criteria, the five following voting groups may be observed:

1) States with *high per capita voting* for farmer-labor parties
2) States with *high, but declining per capita voting*
3) States with *increasing per capita voting*
4) States with *moderate, but consistent voting*
5) States with *little or no voting*

These five groups of states are now analyzed in turn, in an effort to cast further light on the voting for farmer and labor party candidates in presidential elections.[1]

1. *Area of Consistently High per Capita Voting for Farmer-Labor Parties.*—The states of North and South Dakota and Washington gave high percentages of their Presidential vote to candidates of farmer-labor parties in both 1892 and 1924. But despite these political similarities, the three states differ sharply in many ways.

Both North and South Dakota are dependent upon wheat for their prosperity. They include the largest spring wheat

[1] The charts in this chapter (Charts IV–VIII) are based on these sources: *McPherson's Handbook of Politics, Tribune Almanac, American Almanac,* E. E. Robinson, *The Presidential Vote, 1896–1932* (Palo Alto: Stanford University Press, 1934), E. E. Robinson, *They Voted for Roosevelt* (Palo Alto: Stanford University Press, 1947). The mathematical computations have been made by the authors, who are, of course, responsible for any inaccuracies or errors which may exist. In relating the voting behavior of the various electoral areas to the geographic, climatic, and occupational characteristics of such areas, the authors have had recourse, among others, to the following works: *Statistical Abstract of the United States, 1947* (Washington: Government Printing Office, 1948); Charles O. Paullin, *Atlas of the Historical Geography of the United States* (Carnegie Institution of Washington and the American Geographical Society of New York: 1932); Harold H. McCarty, *The Geographic Basis of American Economic Life* (New York: Harper, 1940). In grouping the various areas according to electoral behavior the authors have adapted several of the techniques found in two pioneering works of Louis H. Bean, *Ballot Behavior* (Washington: Public Affairs Press, 1940), and *How to Predict Elections* (New York: Knopf, 1948).

GEOGRAPHIC PATTERNS 55

region of the country. But South Dakota is also dependent upon livestock and mining. The general political background of the two Dakotas stands in sharp contrast to their dependence upon agriculture. North Dakota maintains a state-owned bank, state-owned flour mills, and state-owned grain elevators. An elaborate insurance system, run by the state, protects farmers against fire, hail, and storms.

CHART IV

STATES WITH HIGH PER CAPITA VOTING FOR FARMER AND LABOR PARTIES IN PRESIDENTIAL ELECTIONS, 1872–1948

(Farmer and labor vote expressed as per cent of total state vote.)

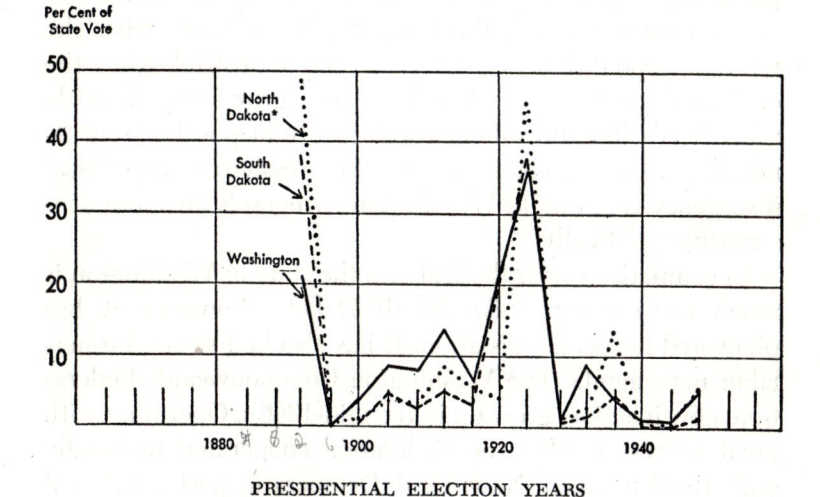

PRESIDENTIAL ELECTION YEARS

* Populists fused with the Democrats in 1892 in North Dakota. The combined vote is here included.

Recent political history in North Dakota is for the most part the story of the Non-Partisan League, founded in 1915 by A. C. Townley. For three decades the League has put into effect a policy of agrarian "socialism." It has done this by operating within the framework of Republican party primaries. The initial strength of the League has been attributed by one authority, James E. Boyle, to a single factor—depressed prices for hard spring wheat.[2] When wheat prices were high, as was the case after the beginning of the World War II boom, the League tended to be relatively weak.

Although in depressed agricultural periods it has followed a similar trend in protest voting to that of its sister state to the north, South Dakota has usually been the more conservative state. The Non-Partisan League has rarely shown much strength in South Dakota.

Washington is similar to the Dakotas in that much of its prosperity depends upon agriculture. Yet the dissimilarities are even greater. Inside Washington great contrasts exist. The western slope of the Cascades is largely industrial, while the eastern slope is dominated by wheat growing and cattle raising. Washington differs considerably in politics from its neighbor, Oregon. The tradition of Progressivism and political insurgency is strong in Washington; Oregon is far more conservative politically.

In comparison with the Dakotas the state of Washington is much more urban. Like North Dakota, however, it has pioneered in social legislation. It has also had strong farmer-labor movements. The Washington Commonwealth Federation, now inactive, grew up in the mid-1930's. Operating with great success locally, the Federation cooperated nationally with the Roosevelt New Deal Democratic party. Internal schisms, over the issue of communism, nearly destroyed the organization in 1946.

The economic dissimilarities between the Dakotas and Washington are great; the same may be said of their politics. Yet, as is illustrated in Chart IV, in times of agricultural de-

[2] James E. Boyle, "The Agrarian Movement in the Northwest," *American Economic Review*, VIII, No. 3 (September, 1918), 505–521.

GEOGRAPHIC PATTERNS

pression they followed similar paths in the support given farmer-labor parties.

2. *Area of High but Declining per Capita Voting.*—The states of Colorado, Wyoming, Nevada, Idaho, Kansas, and Nebraska gave 40 per cent or more of their total vote to Weaver in 1892. Since then, political discontent as expressed by voting for farmer-labor parties has been less intense, although these states supported LaFollette in 1924 more warmly than did many other states further to the east.

These states are younger than those to the east of the Mississippi and, in general, have certain physical similarities. Most of the territory lies to the west of the 100th meridian. The growing period tends to be relatively short. In comparison with other sections of the country, the average rainfall is low. Several of these states, especially those in the Rocky Mountain area, have considerable mining industries.

Utah, Arizona, and New Mexico have been added to this group because their pattern of voting for farmer-labor parties after admission to the Union is most nearly in accord with that of this group of states. Despite their proximity, very important dissimilarities exist in the background of the three states. For instance, Mormonism has been an important factor in Utah's cultural history, as has Catholicism in Spanish New Mexico. Significant differences in the economy of the three states also exist. The surprising thing is that the three states show so similar a voting trend in their support for farmer-labor parties.

Oregon has been included in the states of Group 2, since its voting behavior parallels that of the other states of the group. Yet economically, culturally, and socially, Oregon differs considerably from any other state listed here.

Generally, there has been a slight trend toward decreased voting for farmer-labor parties in Group 2. This trend has been

Chart V (1)
States with High but Declining Per Capita Voting for Farmer and Labor Parties in Presidential Elections

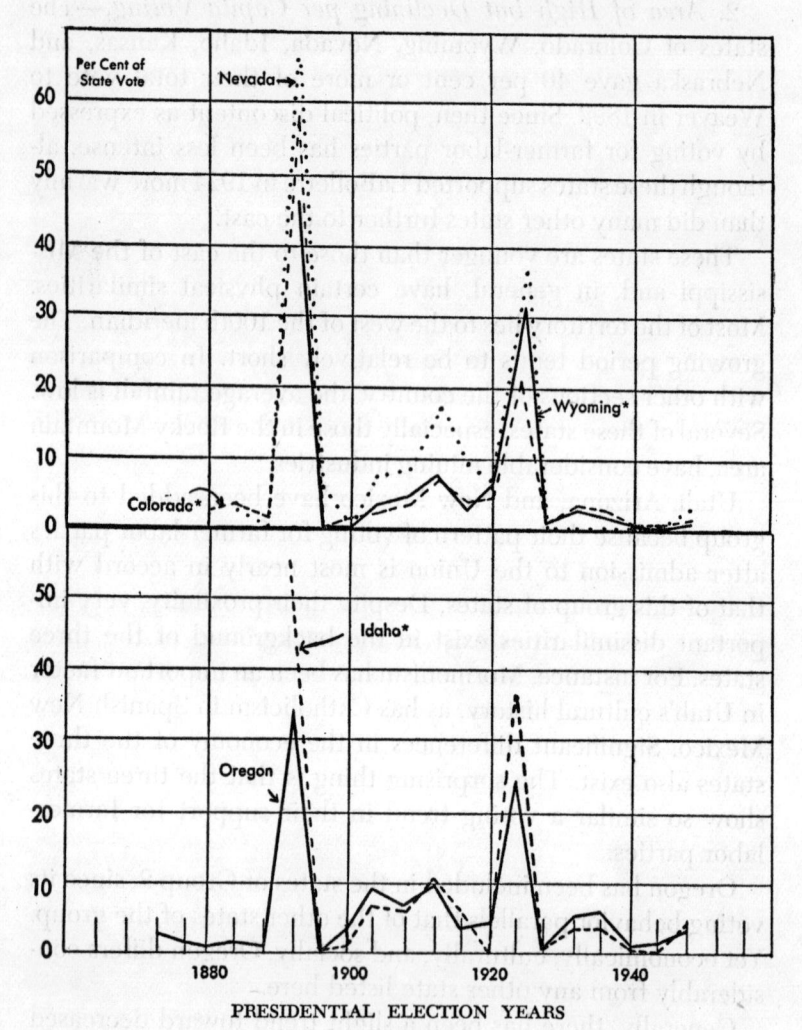

*Democratic-Populist fusion vote for 1892 shown for Wyoming, Colorado, and Idaho.

Chart V (2)

States with High but Declining Per Capita Voting for Farmer and Labor Parties in Presidential Elections

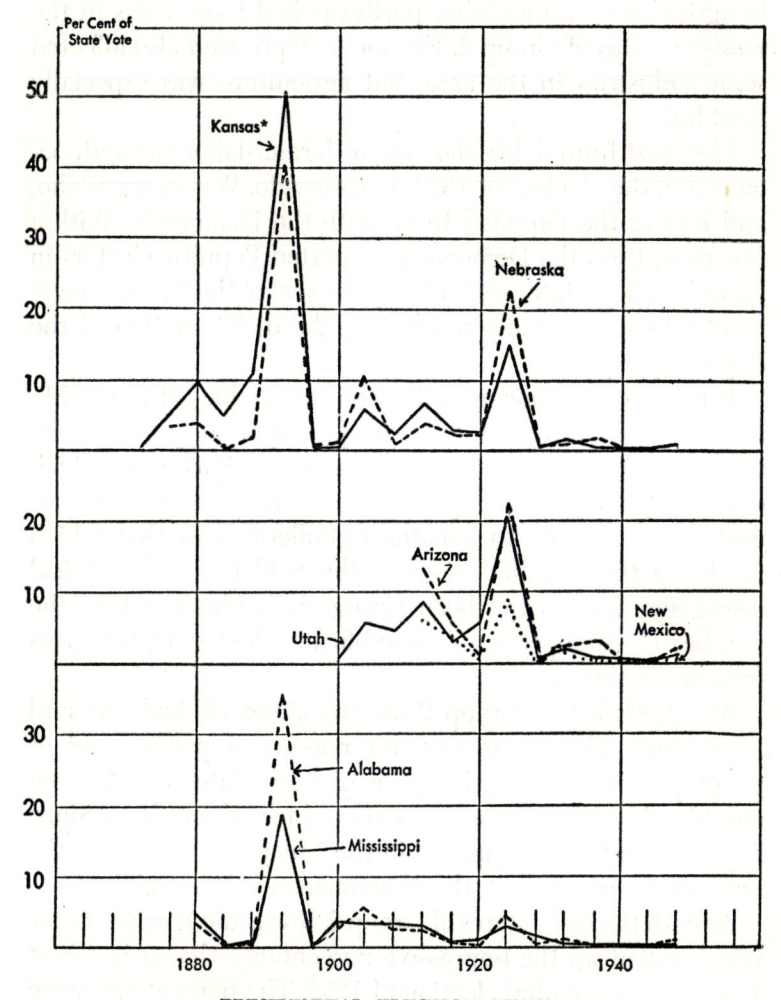

PRESIDENTIAL ELECTION YEARS

* In 1892 the Populists fused with the Democrats in Kansas. The combined vote is here included.

particularly noticeable in recent years in Kansas and Nebraska, once seats of Populist strength.

In 1892 and in 1924, both periods of economic depression in agriculture, farmer-labor parties polled large votes in the western states of Group 2. Economic depression also affected some industries in the area, but agriculture was especially hard hit.

The most formidable showing of farmer-labor strength occurred in the election of 1892. In Colorado, Wyoming, Idaho, and Kansas the Populists fused with the Democrats. With a few exceptions the Democrats supported Populist electors in Nevada. These states were clearly the seat of the Fusion movement which eventually resulted in the disintegration of the Populist party.

It is from this same area that the "farm bloc" of the 1920's and 1930's in Congress drew its principal support, usually under the label of the Republican party. This area, which earlier supported the Populists and later gave a large percentage of its vote to Senator LaFollette, was that which produced the so-called "sons of the wild jackass." Political insurgency has always been strong, as is illustrated by the number of "independent" Republicans elected to Congress from these states.

Also included in Group 2 are the states of Alabama and Mississippi. The inclusion is for reasons of politics, not of geography. The voting behavior of both states has differed considerably from that of the states which adjoin them. Since Alabama and Mississippi are in the "deep" South and in the heart of the cotton belt, the phenomenon is interesting.

As with the other states of Group 2, it was the Populist party which rolled up the impressive percentages shown in Chart V for the Presidential election of 1892. The percentages were 36.5 for Alabama and 19.4 for Mississippi. Populism in Alabama was strongest in the predominantly white and rural counties.[3] The vote given the Populists in both states is the more impressive since fusion did not occur. In the 1892 election the Republican party received only 3.9 per cent of the state vote in Alabama and only 2.6 per cent in Mississippi.

The support for a farmer-labor party, so striking in 1892, has not been maintained in Alabama and Mississippi in recent years. In 1924 the LaFollette vote in these states was only a small fraction of the vote given the Wisconsin Senator in most of the other states which have been included in Group 2. It is likely that enforcement of literacy and poll taxes, white primaries, and the reestablishment of what is ordinarily a one-party political system have contributed to the failure of farmer-labor parties to do well in Alabama and Mississippi since 1892.

3. *Area of Increasing per Capita Vote.*—In some fifteen states there has been a gradual trend toward increasing interest in farmer-labor parties. Many of the states of this group are those in which the greatest absolute numbers of votes have been cast for farmer-labor parties. The group includes California, Minnesota, Montana, Wisconsin, Illinois, Iowa, Ohio, New Jersey, Pennsylvania, New York, Indiana, Michigan, Massachusetts, Connecticut, and Maryland.

There are several noticeable points of similarity among many of these states. Some of them are among the most heavily populated states of the Union. They also contain the greatest concentrations of industry and of capital. When attention is turned from industry to agriculture, further similarities are evident. Six states—Minnesota, Wisconsin, Michigan, Pennsylvania, New York, and Massachusetts—are part of the hay and dairy belt. Iowa, Illinois, Indiana, and Ohio are within the corn belt.

A tremendous dairy belt stretches across much of the re-

[3] John B. Clark, *Populism in Alabama* (Auburn: Auburn Printing Co., 1927).

Chart VI (1)
States with Increasing Per Capita Voting for Farmer and Labor Parties in Presidential Elections

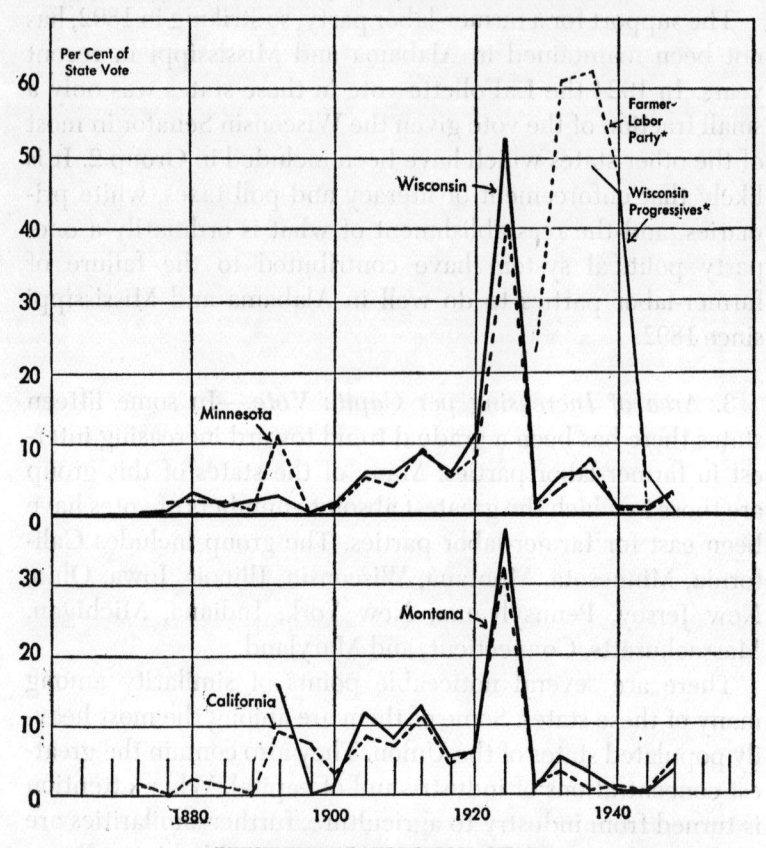

*These figures represent voting for the state tickets of the two parties in Presidential election years. Neither of the parties opposed Roosevelt in these elections.

Chart VI (2)

States with Increasing Per Capita Voting for Farmer and Labor Parties in Presidential Elections

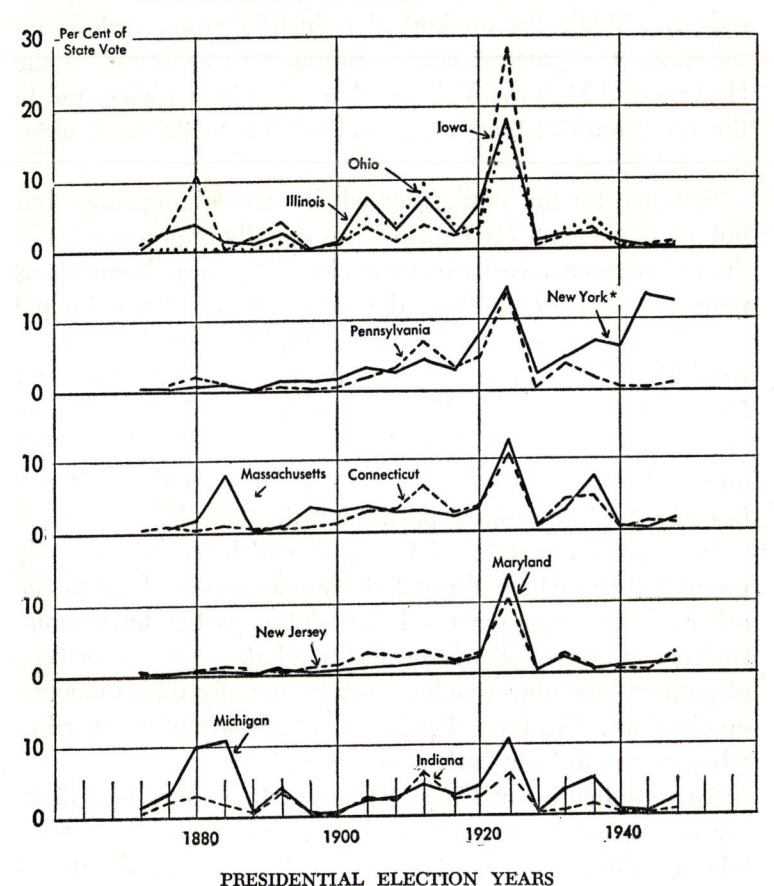

PRESIDENTIAL ELECTION YEARS

* These figures include votes cast for American Labor party and Liberal party Presidential electors.

gion, from central Minnesota and Wisconsin to Pennsylvania and New York. Montana, Minnesota, and Wisconsin have great resources in their forest lands. But probably the greatest common denominator of the states which make up Group 3 is industry. While the amount of industrialization varies considerably, the greatest concentration is to be found in the Hudson and Mohawk Valleys, along the Great Lakes, and in the area from Pittsburgh to Cleveland. This is the heart of industrial America.

So much for the similarities which exist among some, but not all, of the states in Group 3. An equally impressive list of dissimilarities can easily be made, based upon such criteria as percentages of foreign-born, degree of urbanization, principal industries, average rainfall, and geographical location. No one would be likely to contend, for example, that New York and Montana are basically alike, or that Iowa and Pennsylvania are basically alike. Nevertheless, despite all the differences among the many states which make up Group 3, the similarity in voting for farmer-labor parties is quite marked.

In some of the states of Group 3, which are both heavily populated and which alternate between support of one or the other of the major parties, farmer-labor parties have sometimes found it expedient to resort to a balance-of-power type of politics. The objective has been to magnify their influence on elections. This type of politics, particularly with regard to labor parties in New York, is discussed later.

An examination of Chart VI shows that there is a slight preponderance in the percentage of voters supporting farmer-labor candidates in Presidential years in the western states of Group 3. A considerable similarity may also be noted in the voting for farmer-labor parties in the states of this area and the states of Missouri, Kentucky, Tennessee, and West Virginia. These four states lie to the south of most of the states included in the present analysis. In voting behavior, Michigan and

Indiana have historically closely resembled the four Border and Southern states.

The vote shown in Chart VI for Illinois, Iowa, Michigan, and Massachusetts in 1880 was given to the Greenbackers. In all the states of Group 3 the Socialist party carried the principal banner of electoral discontent from the turn of the century until the 1924 LaFollette campaign. Much of the LaFollette vote, of course, was cast under Socialist labels.

A good deal of interest in recent years has been centered on the Minnesota Farmer-Labor party and on the Wisconsin Progressives. These parties are considered in some detail elsewhere in this study. At this point attention is invited to a summary of an examination made by the sociologist George Lundberg of radicalism in the "Northwest." "Northwest" as used by Lundberg referred to North Dakota and to Minnesota. Although North Dakota is not considered in Group 3, the comparison in the Lundberg study between two radical and two conservative communities is pertinent.[4]

Using the vote for candidates sponsored by the Non-Partisan League as the yardstick of political radicalism, Lundberg examined the votes in ten counties of North Dakota and ten counties of Minnesota. In each state he selected five radical and five conservative counties. The elections analyzed were those of 1916, 1918, 1920, and 1922 in North Dakota and 1918, 1920, 1922, and 1924 in Minnesota. Lundberg's findings shed considerable light on political radicalism. First, the conservative counties generally enjoyed more ample rainfall; secondly, they were older and more settled communities; thirdly, they had a much higher percentage of native-born; fourthly, they had uniformly superior economic conditions. Lundberg noted that the most striking feature to emerge from

[4] George A. Lundberg, "The Demographic and Economic Basis of Political Radicalism and Conservatism," *American Journal of Sociology*, XXXII, No. 5 (March, 1927), 719–732.

his study was the presence or absence of foreign-born. He concluded that the Scandinavians and Russians of the radical counties had a cultural background of cooperative enterprise which seemed alien to the native-born farmers of the conservative counties.

Similar detailed studies have not been made for other sections of the United States, but certain features of the Lundberg examination are suggestive. The Socialist strongholds in Milwaukee and in New York City both contained large groups of foreign-born to whom the idea of social democracy was to some degree appealing.

One word of warning is in order on the states which comprise Group 3. The states placed in this region are generally, but by no means altogether, adjacent and contiguous. The presence of California on Chart VI is explainable not by geographical considerations but by political reactions. Nevertheless, it is well to emphasize that the greater number of states in Group 3 are located in the Great Lakes and Middle Atlantic areas. This is also the area, as has been previously pointed out, where the greatest absolute vote has been given to candidates of farmer-labor parties in Presidential elections.

Since the Minnesota Farmer-Labor party and the Wisconsin Progressives did not run Presidential candidates, the vote cast for the state tickets of those parties in Presidential years is shown. The combined farmer-labor Presidential vote was far below the total amassed by the two great parties which ran state but not Presidential slates. The vote shown in Chart VI for New York State includes, of course, the totals for the ALP and the Liberal party. With few exceptions, the patterns shown for the states of Group 3 in previous elections were maintained in the 1948 Presidential election. New York cast 12.6 per cent of its vote for the candidates endorsed by farmer or labor parties, while California cast 4.8 per cent of its vote for the same candidates.

GEOGRAPHIC PATTERNS

4. *Area of Moderate but Consistent Voting.*—A group of Southern states, including Texas, Oklahoma, Louisiana, Arkansas, Georgia, and Florida, have shown consistency in supporting farmer-labor parties in the period since 1872. This support has not been so great as that of the states of the Western area comprising Group 1. The Southern states in Group 4 have rarely missed an opportunity to express discontent, despite the fact that the prevalent one-party system makes it

CHART VII

STATES WITH MODERATE BUT CONSISTENT VOTING FOR FARMER AND LABOR PARTIES IN PRESIDENTIAL ELECTIONS

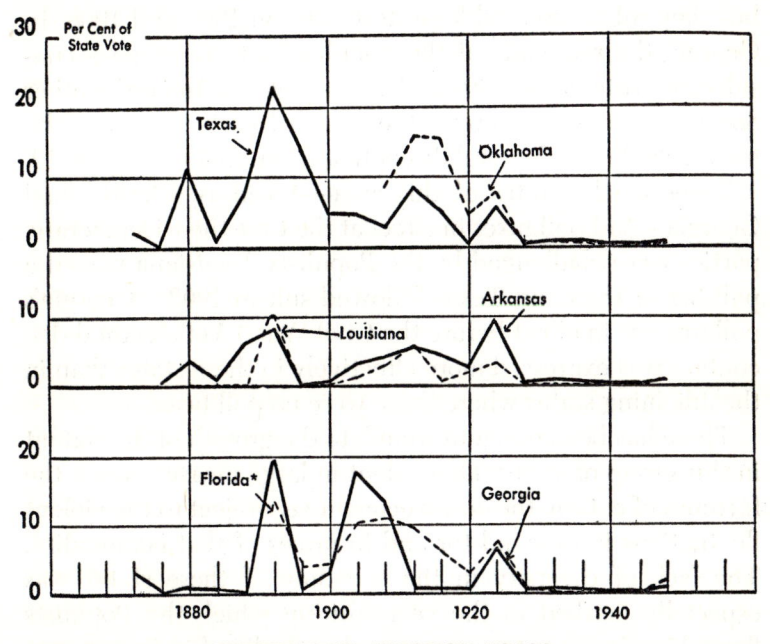

PRESIDENTIAL ELECTION YEARS

* In 1892, Populists fused with the Republicans in Florida. The combined vote is included here.

difficult for such discontent to be registered at all. In fact it was not until 1928, a year promising a certain amount of stability in the agricultural economy, that protest voting for farmer-labor parties practically disappeared in this group of states. After 1932 the Roosevelt Democrats, here as in other states, apparently absorbed the discontent which followed the economic collapse.

With the exception of Florida, these states are all in the cotton belt and all reflected the economic pinch of the 1880's. The depressed economic level in 1892, 1912, and 1924 was likewise reflected in the voting behavior of the area. Georgia and Florida did not express discontent at the polls in the 1880's, but they rolled up sizable protest votes in 1904 and 1908. In Georgia the remnants of the Populist party received a considerable vote in the 1904 and 1908 elections. In all the other Southern states of Group 4 the Socialist party served as the main vehicle for electoral discontent from 1904 through 1912.

Three of the states of this area—Arkansas, Florida, and Georgia—had poll taxes in effect at the time their Democratic parties were challenged by the Populists. Louisiana passed a poll tax in 1898 and Texas followed suit in 1902. (Georgia's poll tax was in effect before the Civil War.) Yet electoral discontent was expressed more effectively in these states than in the adjoining states where there were no poll taxes.

There has been much to stimulate the growth of discontent in this group of states. Dependent in large measure upon the fortunes of cotton, the entire economy is subject to the violent fluctuations in demand for and in prices of that commodity. The lack of diversity in the economy of these states was especially marked in the years during which the Populists flourished. In recent years cotton has declined in importance in the more easterly of the Southern states. Extreme poverty has characterized great numbers of those farming the land.

GEOGRAPHIC PATTERNS

Farm tenancy, always high in this area, has been on the increase. In 1930 the percentages of all farmers who were tenant farmers stood as follows: Texas, 60.9; Oklahoma, 61.5; Louisiana, 66.6; Arkansas, 63.0; Georgia, 68.2; and Florida, 28.4. Mississippi, which is examined elsewhere, led the nation with a percentage of 72.2. The lowest percentages as of 1930 were found in New Hampshire and in Maine, where the figures were 5.3 and 4.5 respectively.[5]

One of the best studies yet made of political behavior inside any state is that of Alex M. Arnett, who examined the Populist movement in Georgia. He found that the Populist vote was as a rule light in cities and towns and much heavier in rural districts. In the region of great plantations, with large numbers of Negroes, the whites tended to band together in support of the Democratic party. In those rural regions where the Negroes were felt not to be potentially a political group of major importance, the tendency was for the poorer white farmers to vote Populist. Arnett found, however, that the issue of white supremacy took precedence over economic issues. This contributed to the dissolution of the Populist party in that state.[6]

The Populist party was also the champion of the poor man in Texas. A study of the movement in the Lone Star state showed that the People's party was socially and economically the party of the poor, small farmer. Some support was given by sheep ranchmen, and most workingmen were sympathetic to the movement. The professional and merchant classes were extremely anti-Populist.[7]

Despite the presence of the inflammatory race issue in the

[5] Paullin, *op. cit.*, Plate 146, Map Q.
[6] Alex M. Arnett, *The Populist Movement in Georgia* (New York: Columbia University Press, 1922), p. 184.
[7] Roscoe C. Martin, *The People's Party in Texas* (University of Texas Bulletin, No. 3308, February 22, 1933), p. 254.

politics of the states which make up Group 4, the voting for farmer and labor parties in Presidential elections during the period from 1872 to 1924 was significant and consistent. In 1892, 1912, and 1924 these states rolled up sizable votes, when expressed on a per capita voting basis, for Weaver, Debs, and LaFollette, respectively.

At various times the political discontent in many of the states of Group 4, as well as in other groups, has been in fact greater than the size of the vote for protest parties. While this has frequently been the case in other parts of the country, it was particularly noticeable in the states of Group 4 in and shortly after the days of the Populists. Thus, as an example, Jeff Davis, who was elected governor of Arkansas as a Democrat from 1901 to 1907 and who later served in the Senate until 1913, might be considered a spiritual heir of the Populists.

In this same group of states there has for some time been a marked tendency for flamboyant orators, allegedly friends of the common man, to assume political power. Jeff Davis, in Arkansas, posed as the friend of the "red necks"; in Georgia, Talmadge father and Talmadge son both sported red suspenders and wooed the rural vote; and in Louisiana the regime begun by Huey Long has shown an ability to maintain itself on a family dynasty basis.

Protest voting of this type, whether it be for good or for evil, has normally occurred within the two-party system and hence is beyond the scope of this study. What can be measured, in the states of Group 4 and in the other groups examined, is the electoral success of those candidates who have run as nominees of farmer and labor parties. Since the days of the Progressives of 1924, however, little voting of this sort has occurred. In 1948 the only state of this group to record an upswing in support for farmer and labor parties was Florida. Nearly 12,000 votes were cast for Wallace, some 2 per cent of the vote of that state.

5. *Area with Little or No Voting for Farmer-Labor Parties.*
—A group of states in New England and in the central South form the last voting area under consideration. In these states there has been little or no voting for farmer-labor parties.

The states of Maine, Vermont, and New Hampshire have much in common and are frequently discussed collectively. Rhode Island, though much more heavily industrialized than the other New England states of this group, has followed the same general voting pattern. Chart VIII shows that only Maine and Vermont of this group of four states marshaled even 5 per cent of the vote in 1924 to the support of LaFollette. The upswing recorded for Rhode Island in 1936 is accounted for by the vote given the Lemke Union party. Even this is inconsiderable when compared with that of many states outside Group 5.

Chart VIII also shows that Virginia, Delaware, North and South Carolina, Tennessee, West Virginia, Kentucky, and Missouri fit into the political area of consistently low voting for farmer-labor parties.

In contrast to the New England states, some of the Southern states of Group 5 show a somewhat greater interest in farmer-labor parties. This is particularly true of North Carolina, which in 1892 gave more than 15 per cent of its vote to the Populists.

Virginia, Tennessee, West Virginia, Kentucky, and Missouri correspond roughly to the corn and winter wheat belt. Four of the states in the group—Kentucky, Tennessee, North and South Carolina—are principal growers of tobacco.

Some of these states—Delaware, Kentucky, and Missouri— were "border" states at the time of the Civil War. Their political alignment has usually been conservative. An exception is the Granger and Populist activity in Missouri during the last century.

CHART VIII
STATES WITH LITTLE OR NO VOTING FOR FARMER AND LABOR PARTIES IN PRESIDENTIAL ELECTIONS

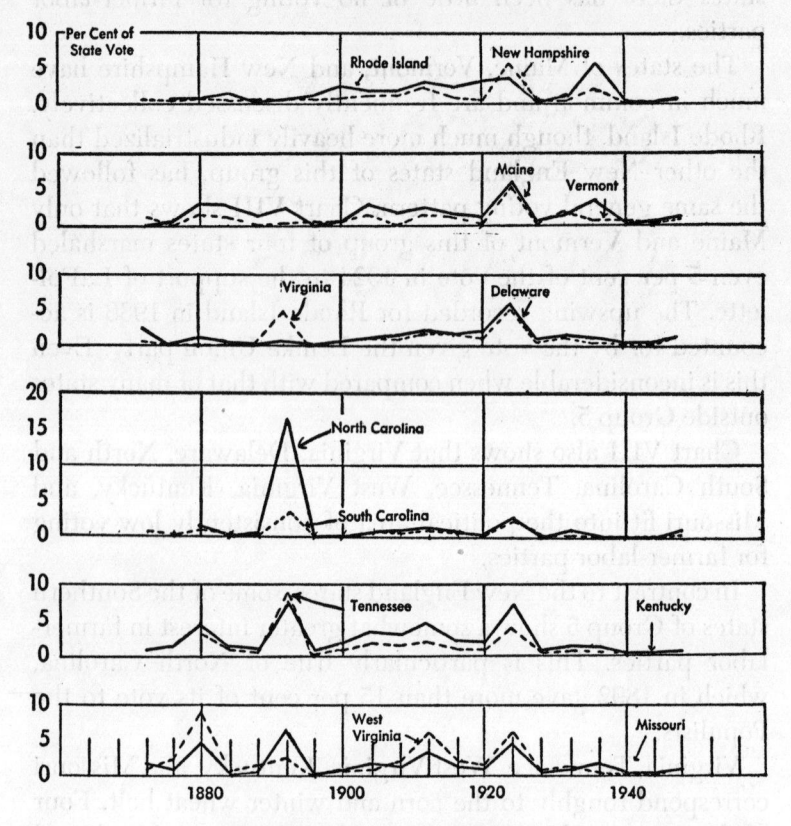

PRESIDENTIAL ELECTION YEARS

In several of these states protest voting has taken the form of fusion with the Republicans. Thus the Virginia Readjusters, when elected to Congress, sat with and supported the Republican party in Congressional divisions. The failure of the Readjuster movement to last has been attributed to the use

GEOGRAPHIC PATTERNS

made by William Mahone and his associates of Negro votes.[8] By capitalizing on the race question the Democratic party was eventually able to drive the Readjusters out of existence. The Democrats then proceeded to establish a single-party system in Virginia. During the 1892 election the Populists in Virginia drew heavily upon the ranks of tenant farmers. These farmers were concentrated in the seventeen counties of the Southside, which constituted the black belt and was economically dependent upon tobacco. But the Populist party crumbled before the race issue.[9]

The history of Republican-Populist fusion in North Carolina ended on a similar note. The injection of racism into the elections of 1896 and 1898 decisively defeated the coalition.

Lastly, it should be pointed out that the pattern of voting in certain states of Group 5—West Virginia, Kentucky, and Missouri—shows many of the peaks that have characterized the voting for farmer-labor candidates in the cotton states. This is not surprising, since political regionalism is a matter of degree, not of absolutes.

SUMMARY OF TRENDS SINCE 1900

From the analysis at the beginning of this chapter it might appear that the center of voting for farmer and labor parties has shifted to the East coast. Such a conclusion, however, is only partially warranted. In terms of percentages of votes cast for Presidential electors of farmer and labor parties, several facts stand out. In 1876 the Greenback party received its biggest percentages in Kansas, Nebraska, Illinois, and Iowa. Centers of the Greenback vote in 1880 were Texas, Iowa, Michigan, Kansas, and Missouri; the highest percentages in 1884 came from Michigan and Kansas. Populist strength in

[8] Allen W. Moger, "The Origin of the Democratic Machine in Virginia," *The Journal of Southern History*, VIII, No. 2 (May, 1942), 183–209.
[9] William Du Bose Sheldon, *Populism in the Old Dominion* (Princeton: Princeton University Press, 1935), p. 92.

terms of percentages was widely spread throughout the area to the west of the Mississippi, although considerable strength existed in Mississippi, in Alabama, and in North Carolina.

After 1900 the states which gave highest percentages of their Presidential vote to electors of farmer and labor parties were notably not those which had been the Populist strongholds. The Socialist vote in 1912 was heaviest in Oklahoma, Nevada, and the area to the west of the Rocky Mountains. In the South Atlantic states it was negligible, except for Florida. By 1920 the new patterns began to emerge even more clearly. Christensen's vote was heaviest in Washington and South Dakota; Socialist strength was fairly strong in New York. With the exception of 1924, when LaFollette's principal support came from west of the Mississippi, farmer-labor strength tended after 1920 to center in North Dakota, Minnesota, Wisconsin, Michigan, Ohio, New York, and Massachusetts. To a very great extent, but with some exceptions, protest voting was heaviest in the states which ring the Great Lakes and in those states of the industrial East.

Analysis of recent Presidential elections alone might lead to the impression that protest voting has shifted entirely to the East coast. Such an impression would be erroneous on two counts. First, the Minnesota Farmer-Labor, the Wisconsin Progressive, and the North Dakota Republican (dominated by the Non-Partisan League) parties did not run independent candidates for President. Instead, they supported the nominees of other parties. The unusual arrangement in New York State whereby a candidate may receive endorsement by a major and by minor parties tends to exaggerate the amount of protest voting in that state. In any event, the American Labor party in 1940, and the Labor party and the Liberal party in 1944, supported Roosevelt. Such a bookkeeping device was not available to measure in North Dakota, Wisconsin, and Minnesota the extent of electoral support given President

Roosevelt by the farmer and labor parties of those areas. This support was indeed considerable.

Secondly, the previous analysis of Congressional and state-wide elections shows that the present parties in New York State have not to date approached their counterparts in Wisconsin and Minnesota in electoral success. The New York parties have never carried the state, never elected a governor, never elected a Senator. When comparisons are made for the state-wide elections of 1934, 1938, and 1942, the Wisconsin and Minnesota parties revealed more strength than the American Labor party has so far shown. Judging solely by the results of recent elections, the center of protest voting in Congressional and state elections has indeed shifted to the East coast. But it seems more than likely that Wisconsin and Minnesota will see revived farmer and labor movements in the future. The Farmers' Non-Partisan League is still alive in North Dakota, and the Minnesota Farmer-Labor party, though officially allied with the Democrats, continues to have a formidable following.

CONCLUSIONS

The preceding statistics show that voting for farmer-labor parties has been more than a series of sporadic phenomena, unrelated in time or environment. For some time there has been a trend for the center of discontent—as expressed in absolute numbers of votes for farmer-labor parties—to follow the general population trend from rural to urban areas. In the case of some of the large industrial states, which are also "doubtful" states in Presidential elections, the implications point to a balance-of-power type of politics by farmer and labor parties in the future.

When the forty-eight states are grouped on the basis of their shares of votes given farmer-labor parties, it is evident that frequently other than voting similarities are also present.

The groupings of states often contain certain likenesses of environment, economy, and climate. Yet in some cases the differences in background in many states which have voted similarly are striking. The states of Northern New England and of the East South Central area have widely differing economies, but their voting behavior has followed a similar pattern. It should also be noted that in some cases states which are close to one another but which have differing cultural backgrounds have followed the same general pattern. Utah and New Mexico are a case in point.

It is difficult to isolate any single cause which fosters the birth and growth of farmer-labor parties. Economic discontent is clearly one factor. Others are an internal struggle for power, ratio of native-born to foreign-born, amount of rainfall,[10] occupational trends, leadership, and racism. In contrast to many purely workers' parties, farmers' parties have not usually followed any fixed ideology.

It is advisable to recall that the introduction of foreign-policy questions on a large scale into Presidential elections complicates the problem of examining conditions under which farmer and labor parties have done well at the polls. With the possible exception of the Socialist party during the years of the First World War, the dominant issues placed before the voters by the more important farmer and labor parties have been domestic. The Wallace movement of 1948 was by no means unique among parties of this type in stressing issues of foreign policy, but the procedure has been, to say the least, unusual. In the past the basic issues raised by such parties have, by and large, related to the national economy. Like its

[10] John D. Barnhart in "Rainfall and the Populist Party in Nebraska," *American Political Science Review*, XIX, No. 3 (August, 1925), 527–540, found that the Populist vote in Nebraska tended to increase in areas which had received less than normal rainfall. Economic, political, and social conditions, which had previously been bad, were made worse by the drought of 1890.

predecessors, the Wallace movement also gave considerable attention to domestic problems.

In the course of this study it has frequently been asserted that, from the voting statistics, some degree of relationship is apparent between the condition of the economy and the size of the farmer-labor vote. It should be borne in mind that the condition of an economy such as ours is a vastly complicated affair. While one section is prosperous, another may be relatively depressed. Inside a given section one industry may be flourishing, while another is decaying. The general relationship between depression and prosperity on the one hand and the voting for farmer and labor parties on the other forms the subject matter of the succeeding chapter.

5

THE ECONOMICS OF PROTEST VOTING

FOUR SIGNIFICANT ELECTORAL SHOWINGS have been made by protest parties of the farmer-labor type. Without exception, these elections occurred in the downswing of a price cycle.

Chart IX shows the course of wholesale prices in this country since 1800.[1] It reveals a price structure dominated by four wars. In each case the inflationary heights were attained in the postwar periods.

It is interesting to note that two of these peaks in voting occurred in the initial downward phase of the price cycle. Even the history following the war of 1812 corroborates this pattern. In the period 1828–1838 local workingmen's parties flourished in Eastern urban centers. These parties represented the only significant attempt at independent political action of a farmer-labor type prior to the appearance of the labor reform and agrarian parties in the late 1860's. While they were entirely local and therefore not comparable to the later farmer-labor parties which were able to compete for votes on a state-wide or national scale, they illustrate the general thesis that farmer-labor parties tend to flourish in the downswing of the price cycle. The Greenbackers in 1878 captured 13 per cent of the total vote some ten years after the first price break. The LaFollette Progressives in 1924 got 16 per cent of the total vote only four years after the price drop following the First World War boom. At the bottom of the price troughs, in

[1] Wholesale price index of farm and non-farm prices is the series published currently by the U.S. Dept. of Agriculture, Bureau of Agricultural Economics. It is compiled as follows: 1800 to 1889, from the price series of Warren and Pearson; 1890 to date, from the U.S. Bureau of Labor Statistics wholesale price index on a 1926 base converted to a 1910–14 base. Price ratio figures were computed from this same series.

CHART IX

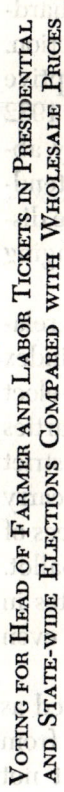

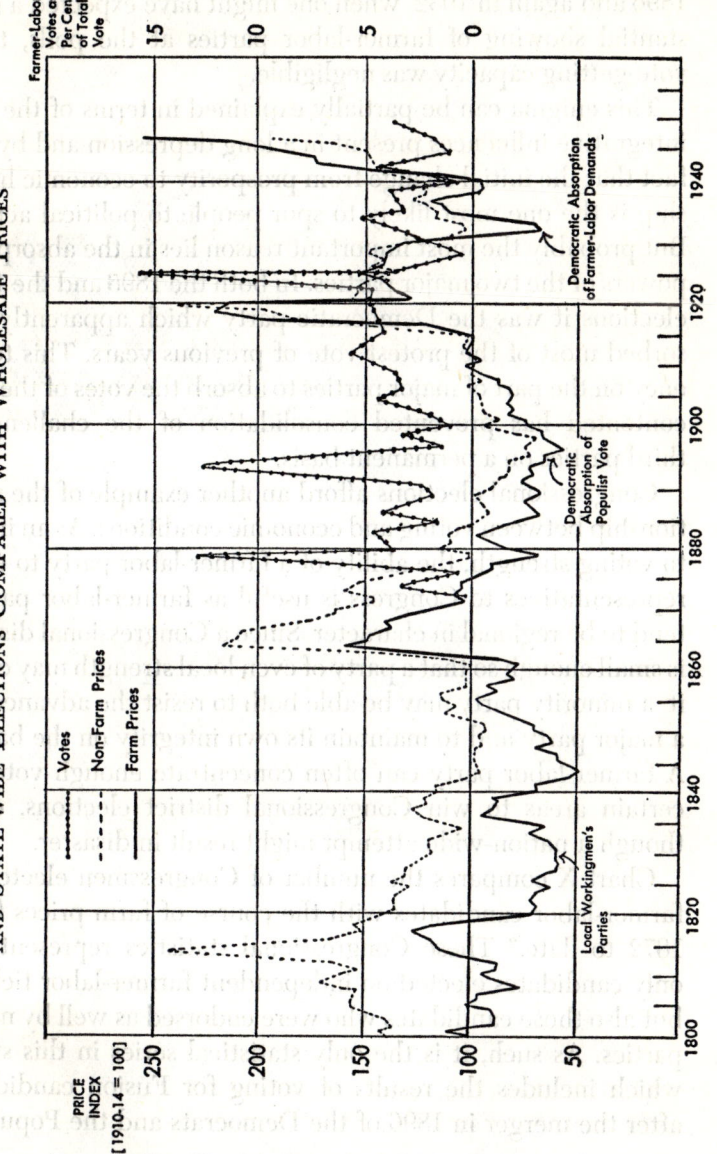

VOTING FOR HEAD OF FARMER AND LABOR TICKETS IN PRESIDENTIAL AND STATE-WIDE ELECTIONS COMPARED WITH WHOLESALE PRICES

1896 and again in 1932, when one might have expected a substantial showing of farmer-labor parties at the polls, their vote-getting capacity was negligible.

This enigma can be partially explained in terms of the disintegrative influences present in a long depression and by the fact that the initial change from prosperity to economic hardship is the one most likely to spur people to political action. But probably the most important reason lies in the absorptive powers of the two major parties. In both the 1896 and the 1932 elections it was the Democratic party which apparently absorbed most of the protest vote of previous years. This tendency on the part of major parties to absorb the votes of the discontented has prevented consolidation of the challenging third parties on a permanent basis.

Congressional elections afford another example of the relationship between voting and economic conditions. As an index to voting strength, the ability of a farmer-labor party to elect representatives to Congress is useful as farmer-labor parties tend to be regional in character. Since a Congressional district is small enough so that a party of even local strength may carry it, a minority party may be able both to resist the advances of a major party and to maintain its own integrity on the ballot. A farmer-labor party can often concentrate enough votes in certain areas to win Congressional district elections, even though a nation-wide attempt might result in disaster.

Chart X compares the number of Congressmen elected as farmer-labor candidates with the course of farm prices from 1872 to date.[2] These Congressional statistics represent not only candidates elected on independent farmer-labor tickets, but also those candidates who were endorsed as well by major parties. As such, it is the only statistical series in this study which includes the results of voting for Fusion candidates after the merger in 1896 of the Democrats and the Populists.

[2] Sources for Charts X and XI are those cited for Chart IX.

CHART X

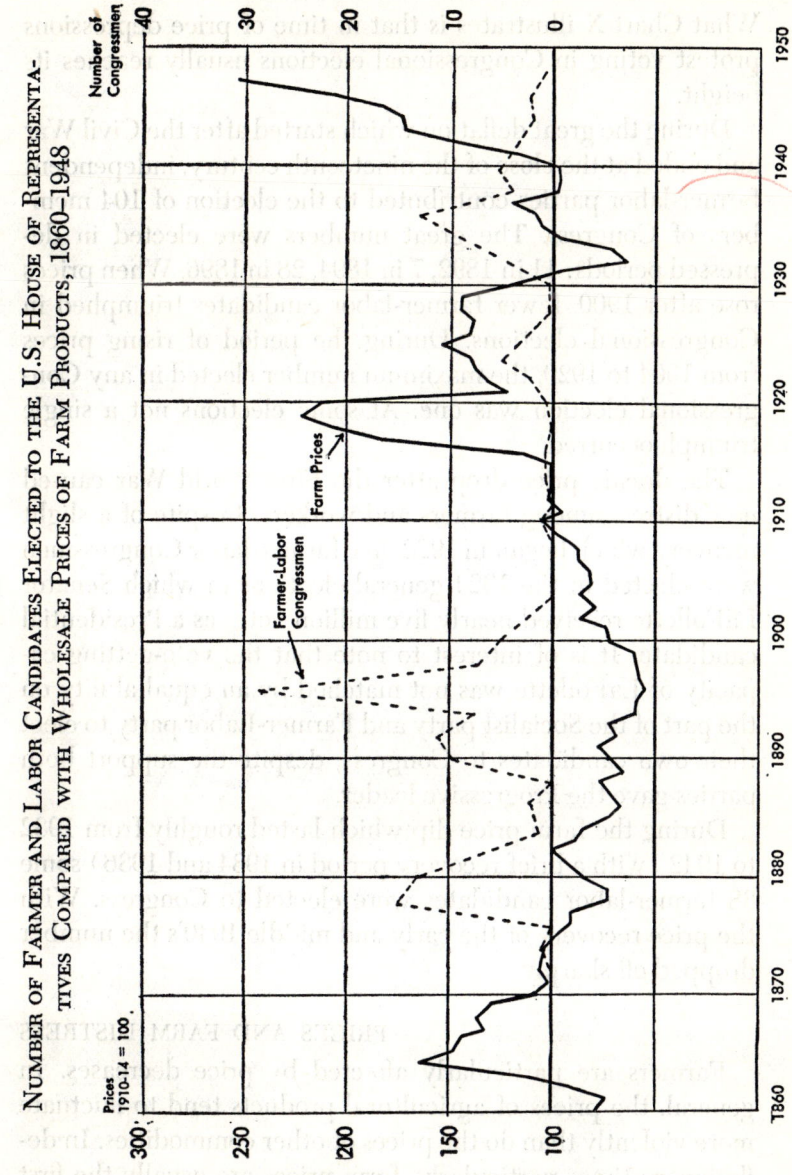

NUMBER OF FARMER AND LABOR CANDIDATES ELECTED TO THE U.S. HOUSE OF REPRESENTATIVES COMPARED WITH WHOLESALE PRICES OF FARM PRODUCTS, 1860–1948

What Chart X illustrates is that in time of price depressions protest voting in Congressional elections usually reaches its height.

During the great deflation which started after the Civil War and ended at the close of the nineteenth century, independent farmer-labor parties contributed to the election of 104 members of Congress. The great numbers were elected in depressed periods: 11 in 1892, 7 in 1894, 28 in 1896. When prices rose after 1900, fewer farmer-labor candidates triumphed in Congressional elections. During the period of rising prices from 1904 to 1920, the maximum number elected in any Congressional election was one. At some elections not a single triumph occurred.

The drastic price drop after the First World War caused new distress among farmers and workers. In spite of a slight recovery which began in 1923, five farmer-labor Congressmen were elected in the 1924 general elections in which Senator LaFollette received nearly five million votes as a Presidential candidate. It is of interest to note that the vote-getting capacity of LaFollette was not matched by an equal ability on the part of the Socialist party and Farmer-Labor party to elect their own candidates to Congress, despite the support both parties gave the Progressive leader.

During the farm price dip which lasted roughly from 1932 to 1942 (with a brief recovery period in 1934 and 1936) some 38 farmer-labor candidates were elected to Congress. With the price recovery of the early and middle 1940's the number dropped off sharply.

PRICES AND FARM DISTRESS

Farmers are particularly affected by price decreases. In general, the prices of agricultural products tend to fluctuate more violently than do the prices of other commodities. In deflationary times particularly, farm prices are usually the first

THE ECONOMICS OF PROTEST VOTING

to fall and tend to drop lower than other prices. Thus the farmer finds himself in the position where the prices he receives for his products are falling faster than the prices of goods which he has to buy.

Chart XI shows this relationship between agricultural and non-agricultural prices and how it varies in timing with the rise and fall in protest voting. The change in the ratio of farm to non-farm prices does not correspond exactly with the change in voting for farmer-labor parties. Yet there are significant instances when they do correspond.

The showings at the polls achieved by farmer-labor parties in 1878 and again in 1924 corresponded in general to a major drop in the ratio of farm to non-farm prices. In particular, after the First World War, the years 1919 to 1925 represented a drastic drop in the relative dollar value of farm produce. It was at this time, in 1924, that the LaFollette Progressives made an excellent showing in several of the Western and predominately agricultural states.

There are also important instances when there appears to be little relationship in timing between these two series. The Populist uprising occurred at a time when the ratio of farm to non-farm prices was more favorable to farmers than it had been during many preceding years. But the agricultural distress was nevertheless widespread and was made all the more serious by the business and price depression which was deepening in the cities. Again, from 1900 to 1915, during the long price rise, farmers were relatively well off, yet voting for farmer-labor parties was on the increase. It was during these years that the Socialist party was picking up the remainder of the discontented Populists in the South and West, but gradually was shifting its actual voting strength to the more urbanized states.

These price indices show how drastic was the financial distress of the farmer during the depression of the 1930's. Yet

CHART XI

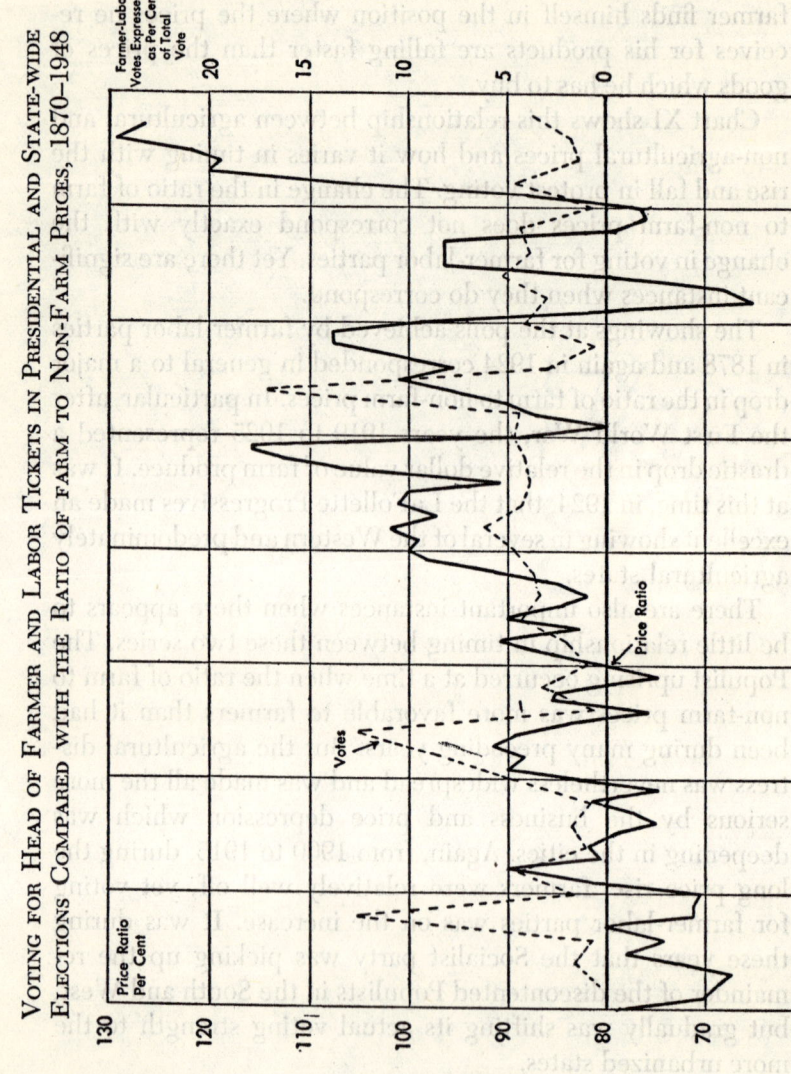

VOTING FOR HEAD OF FARMER AND LABOR TICKETS IN PRESIDENTIAL AND STATE-WIDE ELECTIONS COMPARED WITH THE RATIO OF FARM TO NON-FARM PRICES, 1870–1948

there was little protest voting for independent parties during this period. The Roosevelt Democrats swept into power on a program designed to alleviate hardship among the agricultural and laboring population. After the election of 1932, legislation was passed which consolidated farm support of the administration. The controversial Farm Relief and Inflation Act, creating the Agricultural Adjustment Administration, and the Farm Credit Act were but two of the steps taken in the early years of the New Deal to alleviate the financial distress of the farmers. A leading exponent of extensive reforms for agriculture, Henry A. Wallace, was elected Vice-President under the Democratic banner in 1940. He withdrew from the party in 1947 in protest against the party's foreign and domestic policies to head the ticket of a new farmer-labor party of national stature.

Price deflation has other important implications for the farm economy. In contrast to most of industry, a large part of the cost of agricultural production is represented by fixed costs. Farm prices decline, yet the farmer is unable to reduce his operating expenses sufficiently to balance declining revenues. The burden of fixed costs combined with the declining value of what the farmer produces in relation to the value of other products means that during a period of deflation the farmer operates at a loss. Each successive price drop aggravates the situation.

Distress caused by deflation is further intensified by the fact that many farmers during the upswing of the price cycle tend to invest their profits in land. One result is likely to be a tremendous boom in land values which results in the eventual overexpansion of production in relation to actual demand.

From 1910 to 1923, for example, under the impetus of a rising market, farm mortgage debt increased steadily. The largest increases occurred in 1919 and 1920, the years of the greatest price inflation. The subsequent collapse of the price level left the farmers with large fixed debts and interest charges, while agricultural income fell off sharply. To make matters worse, borrowing by means of mortgages continued for the next three years, this time as distress financing.[3] The heaviness of the burden can be shown by comparing two sets of figures for different years. In 1910 the value of the mortgage debt on owner-operated farms amounted to some 27 per cent of the value; by 1925 the debt amounted to 42 per cent of the value.[4]

During the deflationary period the farmers find it increasingly difficult to meet interest payments. In 1888, it took the sale of 174 bushels of wheat to pay interest of 8 per cent on a mortgage of $2,000; in 1894 it took 320 bushels to make the same payment. According to Solon J. Buck, much of the "agrarian crusade" of the period is explainable in such economic terms.[5]

During the period from 1916 to 1920 the average price paid to farmers for wheat was $1.93 per bushel. At the prevailing rate of interest it required 62 bushels to pay the yearly interest on a mortgage of $2,000. When the average price of wheat fell to $1.11 per bushel during the years 1921–1925, a farmer had to turn over 108 bushels of wheat to cover the same payment.[6] Farmer hostility toward the interests which had originally advanced money through mortgages is easy to understand in terms of the fall in wheat prices following the Civil War and following the First World War.

[3] *Farm Mortgage Credit Facilities in the United States* (Washington: U.S. Dept. of Agriculture Misc. Pub. No. 478, 1942), pp. 2–3.
[4] *Condition of Agriculture in the United States and Measures for Its Improvement*, Report of the Business Men's Commission on Agriculture (National Industrial Conference Board and the Chamber of Commerce of the United States, 1927), Table 7, p. 62.
[5] Solon J. Buck, *The Agrarian Crusade* (New Haven: Yale University Press, 1920), p. 105.
[6] Prices received by farmers are from the regular weighted price averages as compiled by the Bureau of Agricultural Economics, U.S. Dept. of Agriculture.

Another important element in agricultural discontent has been taxes. Historically, the principal levy against agriculture has been the property tax. Since property taxes are calculated on the basis of assessed valuation, they are relatively inflexible and do not fall by any manner of means in a ratio with the drop in the price of farm products.

During 1912, 1913 and 1914, taxes absorbed an average of 11 per cent of total net farm profits, that is, net farm income minus the value of the labor put into production by the farmer and his family. It was reported that taxes collected from all farms during the crop year of 1921 amounted to six times the total net farm profits. At the end of 1922, the tax situation had improved, but taxes still were absorbing two-thirds of the total net agricultural profits.[7]

In addition to the problem of the relatively fixed operating costs and inflexible property taxes, transportation charges have generally troubled farmers. Because of their location at great distances from the principal markets, Southern and Western farmers in particular have had to consider freight rates as an important cost element. It is therefore small wonder that agrarian reformers have habitually attacked high freight rates, railroad stock watering, alleged discrimination in favor of large shippers, and grain elevators operated under monopoly conditions in conjunction with certain railroads.

While railroad regulation finally grew out of the Granger agitation, such regulation did not solve the transportation cost problem to the satisfaction of many farmers. This was especially the case after the First World War. From 1913 to 1917 freight rates on fifty representative agricultural products rose 17 per cent. From 1918 to 1925, during which period a major drop in agricultural prices occurred, the rates on the same products rose an additional 40 per cent. In some cases

[7] *Condition of Agriculture*, p. 81.

the freight charges were greater than the sales receipts for certain products.[8]

The reaction of farmers to freight charges, to taxes, and to interest payments has been essentially that of businessmen. While it may appear that on occasion farmers have charged blindly in their attacks on railroads, elevators, and banks, they have acted understandably. As one historian of early Western agrarian agitation expressed it, the farmers, when viewed in the perspective of our history, act "as ordinary businessmen, slightly over-individualistic perhaps, seeking to correct injustices in the marketing and credit systems, trying to cut down fixed costs which threatened to devour their margin of profit, and endeavoring to build up the wealth of the community of which they were citizens."[9]

WAGES AND VOTING

While a decline in commodity prices has usually corresponded in timing with a rise in the protest vote, the course of wages has shown little or no such connection. Until the First World War wages were less flexible than commodity prices. Money wages did not rise so high as the prices of goods nor did they fall so low in the general price depressions. Of significance in interpreting voting for farmer-labor parties is the fact that during the period of the late 1890's, which was a time of great financial distress for agriculture, money wages hardly declined, while real wages increased.

Again, in the 1920's, when LaFollette attempted to challenge the two major parties, real and money wages were maintained at their wartime levels, while the general commodities price level and especially prices for farm products

---

[8] *Condition of Agriculture*, Table 10, p. 84. Figures are computed from index numbers of railroad freight rates on fifty representative agricultural products as compiled by the Department of Agriculture. See *Yearbook of the Department of Agriculture, 1926*, p. 1248.

[9] Benton H. Wilcox, "An Historical Definition of North Western Radicalism," *Mississippi Valley Historical Review*, XXVI, No. 3 (December, 1939), 394.

were undergoing a considerable deflation. Clearly a decrease in wages is not among the factors which could be used to explain the large vote which Senator LaFollette received in 1924 from both urban centers and agricultural areas. It was not until 1932, at the time of the worst depression to date in American history, that wages took a significant downward trend.

Real wages are, of course, a better measure than money wages of the actual earning power of labor. Measuring as they do the relationship between the amount of money a worker receives and his cost of living, they show his effective purchasing power. With the exception of temporary fluctuations, the trend of real wages has been upward since the start of the 19th century.[10] This gradual increase in the purchasing power of labor's dollar may be one reason why political action to change the *status quo* has appealed to so few in the ranks of labor and why a large segment of organized labor has been satisfied to limit its energies to purely economic instead of political matters.

THE CONDITION OF BUSINESS AND VOTING FOR FARMER AND LABOR PARTIES

Historically, an extended drop in business activity has normally been followed by an increase in the number of votes accorded farmer-labor parties. Chart XII compares the course of business activity with protest voting.[11]

For convenience in analyzing this chart, the time series it depicts can be broken down into three periods. The first, running roughly from 1864 to 1898, occurs during the down-

---

[10] The trend of money wages and of real wages was derived from the composite wage index compiled at the Federal Reserve Bank of New York by Carl Snyder.

[11] Index of business activity is that compiled by the Cleveland Trust Company. Annual figures used here were arrived at by a simple averaging of monthly figures.

CHART XII

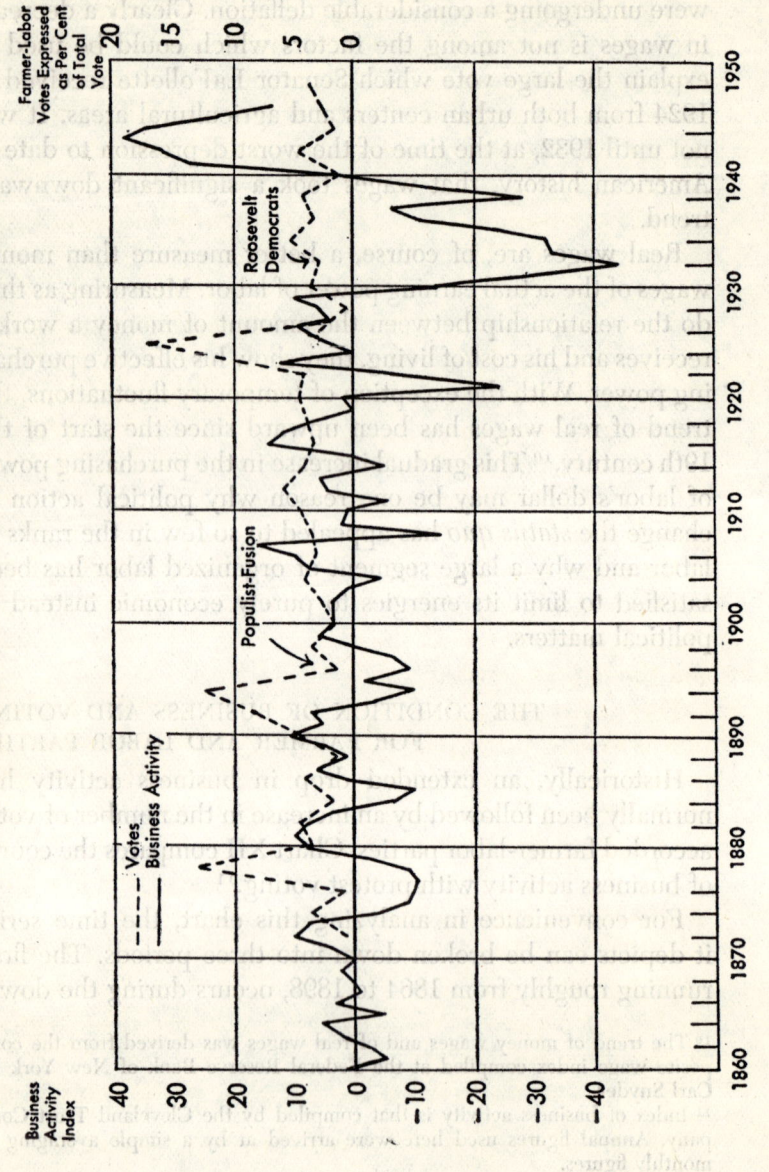

VOTING FOR HEAD OF FARMER AND LABOR TICKETS IN PRESIDENTIAL AND STATE-WIDE ELECTIONS COMPARED WITH BUSINESS ACTIVITY, 1872–1948

swing of the price cycle. Both major drops in business activity occurring during this period were accompanied by a major increase in voting for farmer-labor parties. As was mentioned earlier, in 1896, in the heart of the depression, the ground was cut out from under the Populists by Democratic absorption.

The next period, running roughly from 1898 to the end of the postwar boom in 1920, corresponds with the upswing of the price cycle. During these years, farmer-labor parties at no time were able to attract the votes which their predecessors had previously received. Nor were the depressed periods so serious as those during the great price deflation. As W. L. Thorp pointed out in his classic study, "the relative duration of the prosperous and depressed phases of business cycles is dominated by the secular trend of wholesale prices." By way of historical evidence, he continues: "In the three periods of rising price trends since 1790, the prosperous phases of the cycles have been prolonged and the depressed phases have been brief. In the three periods of declining price trends, the prosperous phases of the cycles have been relatively brief and the depressed phases prolonged." [12]

These observations by Thorp provide a tentative explanation as to why, during the years 1898 to 1920, little significant protest voting occurred. Only the brief fluctuations in 1904–1915 were reflected, in part, by a small peak in voting which the Socialist party attained in 1912.

The third period covers the price collapse after the First World War. In this period, the big vote for LaFollette followed the very sharp and severe drop in business activity of 1921 and 1922. In 1932, the political discontent caused by the biggest depression in our history was apparently absorbed by the Democratic party.

Historically, the relationship between protest voting and

---

[12] Willard Long Thorp, *Business Annals* (New York: National Bureau of Economic Research, 1926), p. 71.

economic conditions would seem to be this: When a major price drop is combined with a severe or prolonged drop in business activity, agrarian and labor protest voting will either reach new heights or else one of the major parties will reach out to absorb this new challenge.

UNION ACTIVITY AND VOTING

A strike is symptomatic of discontent and is an expression of protest. Generally workers resort to this type of economic weapon when they feel they may improve their own economic position. They strike for the most part when their bargaining position is good—when they can reasonably expect to win and not to lose a dispute.

The long term trend in strikes has followed the price curve.[13] Except for the years 1912 to 1915, the period starting with the turn of the present century and ending with the First World War was predominantly an era of business prosperity and expansion. The gain in union membership was tremendous.

The steadily rising cost of living in the years before the First World War had the effect of sparking the newly developing labor movement into calling strikes for higher wages. The objective was to maintain purchasing power. During the period from 1900 to the end of the First World War militancy in striking was in general matched by militancy in voting. But it was not until after the major break in prices had occurred that political action reached its height.

The price fall in the 1920's, coupled with the depression and unemployment of the immediate postwar years, resulted in a deterioration of labor's bargaining position. Depression

---

[13] This summary of the trend of strikes depends on research developed by John I. Griffin in *Strikes, a Study in Quantitative Economics* (New York: Columbia University Press, 1939). This study is based on the strike figures of the Bureau of Labor Statistics. Dr. Griffin, however, by his own methods was able to bridge the statistical gap of 1906–1915, during which years the government did not collect strike statistics.

and the open shop drive carried on at the time by many powerful employers turned labor toward political action as a method of improving its position. It is significant that in 1924 the American Federation of Labor's executive committee abandoned its traditional hands-off policy in Presidential elections and endorsed Senator LaFollette for President. The AFL shift was not, however, caused exclusively by the depression.

With the return of apparent prosperity in the late 1920's the total vote for farmer-labor candidates in all elections decreased. After the first Rooseveltian victory in 1932, the sweeping reform legislation passed by the Congress served as a tremendous boost to organized labor. Under the National Industrial Recovery Act and later the Wagner Act, union membership more than doubled between 1933 and 1937. By the time of our entry into the Second World War union membership had quadrupled.

The strengthening of labor's position combined with a slight price rise beginning in 1933 led to an increase in both the number of strikes and the number of workers on strike.

By 1941, as our economy went on a wartime basis, prices again turned upwards, this time in an inflationary spiral that was to last well beyond the actual war years. In a recurrence of the pattern which existed prior to the First World War, more strikes were called as prices rose, and larger numbers of workers were involved. Again there was a slight rise in political discontent as expressed in the number of votes cast for farmer-labor parties.

This relationship between prices, business activity, and strikes has been well described by E. B. Mittelman in his study of Chicago labor in politics during the period 1877 to 1896. He points out that as prices turn upward from the bottom of the depression, labor

now turns from the demand for shorter hours to a demand for higher wages to meet higher prices. This demand continues as prices continue to soar until finally it recognizes the dilemma of catching up with prices which it itself, in part at least, advances with each increase in purchasing power; which it secures, at times, only at the end of costly strikes. Once [labor] recognizes this dilemma, it goes into politics to curb that element in the community which in the meantime has been profiting from the upward changing prices. . . . It is only the leadership that goes into politics at this point—that is at the crest of the business cycle. The rank and file still have steady employment and are willing to let well enough alone. When the cycle finally takes its downward course and the factories are closed for part time or altogether, they too begin to view politics with favor. . . . Only at the end of a series of strikes aimed at keeping wages abreast of rising prices has labor gone into politics. The masses did not go in until depression came, and went out of it as soon as time brought relief.[14]

While the foregoing rationale of how strikes and protest voting fit into the changing pattern of the American economy implies a relationship between numbers of people on strike and protest voting, it should be remembered that strikes are primarily a symptom and not a cause of economic discontent. They are affected by the basic changes in the economy which in turn influence protest voting.

It is still true, also, that to date the tremendous potential voting strength of organized labor has rarely been made available to farmer and labor parties. Not until 1924, after a marked decline in union membership, did the total number of votes cast for protest parties exceed the total of union membership. This situation has not been duplicated. But from the point of view of the future, the steady growth of union membership which started in the days of the New Deal and which received new impetus during the war years, is of great importance politically.

The membership of all unions in 1924 amounted to some

[14] Edward B. Mittelman, "Chicago Labor in Politics 1877–96," *Journal of Political Economy*, XXVIII, No. 5 (May, 1920), 407, 426. Reprinted by permission of the publisher.

three and one half million persons. Had all of them voted (or been able to vote), they would have represented some 12 per cent of all votes cast for all parties. In 1948 union membership totaled some 15 million.[15] This figure was equivalent to 30 per cent of the votes cast in the Presidential election of that year. While this great potential voting bloc may never express itself through independent parties, it certainly will be wooed by one or both of the major parties.

COMPARISON OF ECONOMIC AND POLITICAL FACTORS

When various of the economic and political factors discussed to this point are drawn together, certain tendencies become evident. The relationship between voting and economic change becomes most clear when specific periods of our economic history are used to provide a framework for political analysis. There is no general agreement on what constitutes a period of "good" times as against a period of "bad" times. Yet specific periods of business cycles stand out by virtue of the great contrasts between certain booms and busts.

The principal depressions since the Civil War were those of 1873–1879, 1883–1886, 1893–1897, 1904, 1908, 1914–1915, 1920–1922, 1929–1940. Several of these depressed periods were brief, notably those of 1908 and 1914, termed the "panic of 1907" and the "rich man's panic" respectively. Such brief depressions appear not to have been reflected in the political sphere. Most of the longer depressions, however, have preceded periods of political activity by farmer and labor parties.

The first post-Civil War depression became evident in May of 1865 and lasted about a year. In November, 1873, began a much more severe depression, which lasted until the fall of

[15] This discussion of union membership is based on the compilations of the U.S. Department of Labor, Bureau of Labor Statistics, covering the period 1897 to 1947.

1879. The panic of 1873 was precipitated by the failure in September of the nation's leading brokerage firm, Jay Cooke & Company. Failures of numerous banks followed, the Stock Exchange was closed for ten days, and there was a partial suspension of specie payments. In July, 1877, the great strikes broke out, to be put down subsequently by the use of troops. This was also the year of the Molly Maguire riots in the coal fields. The early part of 1879 was characterized by inactivity in trade and industry, a record number of business and bank failures, and a decline in stock values. Drawing upon labor discontent in the East and agrarian discontent in the West, the Greenback party spearheaded the attack upon the major parties. This party reached its peak in 1878, when it received 12.7 per cent of all votes cast in the state-wide elections.

The famous panic of 1893 began in July of that year. A brief and modest recovery occurred in the latter part of 1895, which was followed by another depression from January, 1896, through November, 1897. In general, the period from 1893 to 1897 was one of widespread depression, unemployment, and low prices. Wheat and corn prices turned upward in 1897, stock market prices rose, and the year ended with a decided upward trend in most fields. During this period the voting for farmer and labor parties reached another peak. Most of the votes came from agrarian areas in the Midwest and the Southwest.

The farm depression was somewhat longer than the general depressed period of business activity. The year 1892 was not good on the farms although it was relatively prosperous for business and industry. The bottom of the second phase of the 1893–1897 depressed period came in 1896, when intense industrial depression and very low farm prices came at the same time. From 1892 through 1894 the Populists were at the height of their power. Following the McKinley victory in 1896, the

Populists rapidly disappeared. In November, 1897, good times were again coming back.

The years from 1900 to the First World War were marked by a rising price level, general agricultural prosperity, and business expansion. It will be recalled that this was also the period of consolidation of the labor movement. Union membership and strikes both increased. Voting for farmer and labor parties increased in urban centers. The principal champion of agricultural discontent was the Socialist party, which had picked up the pieces left by the disintegration of the Populists in the West and South. The 1913–1915 depression in general coincided with the peak of Socialist strength, which was reached in 1912. The depression was not severe, nor did the Socialists attain the voting strength of the Populists or Greenbackers.

A sharp, severe depression started in October of 1920 and lasted through September, 1922. This depression affected primarily the farmers, for business and industry recovered quickly. Prices for agricultural products enjoyed no such recovery. In the elections of 1920 the number of votes cast for farmer and labor party candidates topped the number cast in the war era. In 1922 the number of such votes sagged slightly, then soared in 1924. The extensive character of the depression-caused hardship undoubtedly contributed substantially to the impressive size of the protest vote given the 1924 Progressives. But clearly, other factors were of importance.

The year 1924, while it followed close on the heels of the sharp business depression of 1920 and 1922, was itself relatively prosperous, except on the farms. The tremendous personal drawing power of Senator LaFollette should not be underestimated. The spirit of unrest which follows every war was probably another factor. The period was one of great economic and psychological uneasiness. Large numbers of radicals were hunted down and sent to jail or deported. Many big employers pressed for the open shop, which they termed the "American Plan." That both railroad unions and farm organizations could support LaFollette is a fact of tremendous significance. The endorsement of the Progressive candidate by the American Federation of Labor's executive council also illustrates the widespread nature of the discontent.

After 1924 a period of business and industrial prosperity set in which convinced many that depressions were a thing of the past. Hardship in agriculture and coal mining remained, but this was widely ignored in the light of the more rosy predictions for the future. The voting for farmer and labor parties dropped drastically in 1926 and dropped again in 1928 to a near-record low. It did not move upward again until 1930.

Meanwhile the most famous depression of our economic history had begun. Beginning with a stock market collapse in the late fall of 1929, the depressed period of business activity lasted, with ups and downs, until mid-1940. At that time the boom caused by the war in Europe was getting into full swing.

The political struggles of the 1930's are in many ways reminiscent of the battles of the 1890's. Despite the intensity and bitterness of the depression, the voting for farmer and labor parties was much less than might have been expected. The share of the total vote cast for farmer-labor parties was smaller, in fact, than it had been during many prosperous periods. The explanation lay in the nature of the New Deal program. Whatever many persons may have thought of New Deal economics, it was true that farmers and union members felt they were receiving many benefits from the Roosevelt administration.

Again, as in the 1890's the two-party system demonstrated tremendous flexibility. Many of the reforms recommended by minor parties were adopted by the Democratic administration and made into the law of the land. Since a good many of the programs of minor parties were directed at alleviating the condition of workers and farmers, it is not surprising that the vigorous New Deal program, contradictory though it may have been, was to so large an extent supported by organized labor and by leading farm organizations. In any event, except for local parties in New York, Wisconsin, and Minnesota, which never contested the leadership of Roosevelt in Presidential elections, voting for farmer and labor parties was at a low ebb.

By the middle of 1940 the depression of the 1930's was rapidly disappearing, to be replaced by a period of unprecedented business and agricultural activity as this country entered World War II. Business activity, though unevenly spread, was general enough to stimulate a very real measure of economic prosperity. Union membership rose to new heights. The most notable characteristic of the period was not the war-induced prosperity but rather the length of the period of high business activity. After the First World War a price break and a depression followed in little more than a year. After the Second World War prices continued to climb, employment was maintained at an all-time high, and even by the time of the 1948 election no sizable over-all price drop had occurred.

On the other hand, during 1948 some signs indicated that the crest of the boom had passed—unemployment rose in some industries, prices fell in a few lines. Some farmers were worried as to whether many of the New Deal measures initiated in the 1930's to allay farm distress would be continued in the event of a Republican victory. In the record of the 80th Congress many farmers found grounds for pessimism on the question of price support for farm products.

Early in 1948 a new farmer-labor party, the Wallace Progressives, was organized to challenge the major parties. As had been the case in the 1920's, the centers of third party activity were in New York, Illinois, California, Minnesota, and Wisconsin. In Illinois, where the Progressives failed to get on the ballot, those persons who wished to express dissatisfaction with the major parties voted for the Socialist or the Socialist Labor candidates. Some 15,000 voters did this.

In general, Wallace's party did poorly at the polls. Agrarian and labor protest voting apparently took the form of supporting President Truman. That the Progressives captured so little support was not surprising. As we have seen, independent protest parties of the farmer and labor type have never done well in periods of general prosperity.

CONCLUSIONS

Certain major conclusions are evident from the foregoing analysis. The first of these is that there is a relationship between prices, wages, unemployment, and economic well-being, on the one hand, and the voting for farmer or labor parties. This relationship is by no means an exact one. Economic man is not so carefully tuned to the times that a ten-point change in a Bureau of Labor Statistics index will result in a ten-point change in voting behavior. Many factors which are not economic in nature vie for the attention of farmer and labor voters. In addition, the interests of farmers and union members are not always identical. When coincidence of interests is present, it is often obscured by traditional antipathies. In many cases the interests of different farm groups are in opposition to each other. This is also true of workers' groups. Thus, to cite only one example, skilled and unskilled workers in the same area may support different candidates at the same election. With all of these qualifications, however, the fact still remains that the vote for farmer and labor parties does respond to the general economic situation.

When has this relationship in the past been the closest? This question leads to the second general conclusion. The relation in the past has been closest when the economy was going downhill, particularly when a declining price level was combined with a business recession but before the absolute bottom was reached. When the absolute bottom was reached, one or both of the major parties recognized the widespread character of the discontent. The rumblings from farmers and workers were heeded. A program of reform and action was set in motion. When this happened, the vote for farmer and labor parties tended to drop off sharply. Thus, the paradox that farmer and labor parties have tended to do most poorly at the very time when an economic determinist might be led to anticipate that they would do best. This phenomenon may not be an eternal feature of American politics; but thus far in our history it has been the rule rather than the exception.

6

STRATEGY AND TACTICS

THE FIRST STRATEGICAL OBJECTIVE of any farmer or labor party is to mobilize for political action the greatest number of farmers or workers. There is thus a fundamental distinction to be drawn in every election in which such parties compete. This distinction is between the actual vote of the parties, as recorded at the polls, and the potential vote which would have occurred if class mobilization had been complete.

In practice it is not possible to achieve any great degree of political homogeneity among groups which are assumed to have at least some large measure of unifying economic interests. This statement is readily verified by a comparison of the number of persons in certain occupations with the votes given selected parties. In 1876, for example, there were about seven million males gainfully employed in agriculture. Many of these persons could not vote because of age and residence requirements. Yet, assuming for the moment that the majority were entitled to vote, it is interesting to compare the number of such males in agriculture with the vote given the Greenback party. Peter Cooper, the Greenback candidate, received only 81,000 votes. Again, in 1936, about thirteen and a half million persons, male and female, were engaged in manufacturing and mechanical industries alone.[1] Yet the vote of all labor parties combined, including that of the ALP in New York, was only a little more than 1,400,000. Class mobilization has thus far in our history been pitifully weak. Most workers and farmers have usually supported the major parties.

[1] For a thorough analysis of census figures relating to occupations, see H. Dewey Anderson and Percy E. Davidson, *Occupational Trends in the United States* (Palo Alto: Stanford University Press, 1940).

STRATEGY AND TACTICS

Another strategical objective of farmer and labor parties is to effect a union of farmer and labor groups into one party. This recurrent hope is based on the belief that the mathematics of the electoral process would add up to the possibility of victory if such a union could be consummated. In its most general terms, the formula would read: All farmers (plus agricultural workers and inhabitants of rural communities dependent upon farm prosperity for their own well-being) combined with all factory and industrial workers would give the thus united party an overwhelming majority anywhere, anytime.

Such a prospect is not, however, within the possibility of attainment under the historical conditions which have thus far existed in America. A more refined formula, with limited objectives, has thus been developed. This is that all organized farmers plus all organized workers plus their sympathizers in other economic groups might do very well indeed if strongly organized into one political party.

Numerous attempts to realize this hope have occurred. We shall limit ourselves to a consideration of two outstanding examples.

When the National party was officially founded at the Toledo conference of February, 1878, the bulk of the delegates present were representatives of farm organizations and of labor unions. The unofficial label of the new party, the Greenback Labor party, gives more than an inkling as to the hopes and aspirations of its founders.

The Toledo resolutions included both agrarian and labor planks, but the *common denominator of the various groups was a demand for inflationary action by the federal government.* From today's perspective the position taken by labor groups at the Toledo conference may not seem understandable. Yet the explanation is quite clear. The story goes back to William Sylvis, a labor leader who played a key role in organizing the National Molders' Union in 1859, and the subsequent National Labor Union formed at the close of the Civil War.[2]

The Union stood for cooperative enterprise, improvement of working conditions for women, better housing, the eight-hour day, and currency reform. It was this last demand which attracted the greatest amount of attention at the time. The Union's position was that the root of all difficulty lay in the "gold disease," and that a program calling for additional paper money, and other money and banking changes, was the solution.

Calling its assorted proposals the "American system of finance," the Union blamed unemployment after the Civil War on falling prices caused by a currency shortage. Withdrawal of money, it was argued, had forced factories to close, or to curtail production, with the result that workers were forced into unemployment. The Union established a National Labor party in 1872, and adherents of this group were subsequently prominent in setting up the Greenback Labor party.

While the Greenback movement was predominantly agrarian in character, it did form many local alliances with state labor reform parties. In Connecticut, New York, and Massachusetts, as well as in the West, the movement made its influence felt in local elections. In 1877 the Greenback and labor reform parties put up joint tickets in Massachusetts and in Ohio.

Another outstanding example of attempted farmer and labor coalition took place in 1924 during Senator Robert M. LaFollette's campaign for the Presidency on an independent ticket. The impetus to the Progressive movement of 1924 came

[2] See Jonathan Grossman, *William Sylvis, Pioneer of American Labor* (New York: Columbia University Press, 1945), for an interesting biography of this early labor leader.

from three sources: the railroad brotherhoods, the LaFollette machine in Wisconsin, and the Non-Partisan League and farmer-labor elements of Minnesota and the Dakotas. It was these groups which directed the nominating convention which met at Cleveland on the Fourth of July, 1924.

When the executive council of the American Federation of Labor endorsed the independent candidacy of Senator LaFollette, the spectre was raised that the AFL membership of 2,865,979 could swing the election in conjunction with other labor and some farmer groups. Acceptance by both the Socialist and the Farmer-Labor parties of the LaFollette candidacy raised further threats to the major parties.

The results of the 1924 election are analyzed elsewhere. At this point it is important to note only one fact: Although Senator LaFollette polled nearly five million votes, the threatened farmer-labor coalition did not fulfill the expectations of many of its backers. In particular, the American Federation of Labor was sorely disappointed in the outcome of the election.

This brings us to the third general objective of farmer and labor parties—the attempt to achieve considerable support from the middle class. There are three ways in which such support has been solicited. First, the parties write broad programs intended to attract white collar workers, small entrepreneurs, and members of the professional groups. Secondly, the parties may nominate a candidate of national stature to run for the Presidency, rather than select purely a farmer or a labor leader. Such was the case with the Populist nomination of General Weaver and the Progressive nomination of Senator LaFollette in 1924. This is to be contrasted with the technique of those parties which do not ordinarily go out of their way to solicit middle class support, for example, the Communists and other extreme groups. The third way in which farmer and labor parties have tried to obtain middle class support is by the formation of working alliances with local and state reform groups.

It is generally understood in such situations that, if elected, the farmer or labor party candidate will concentrate on the minimum, not the maximum, objectives of his party. Thus, in the fall of 1947 the industrial town of Norwalk, Connecticut, in a revolt against a long period of rule by the two major parties, elected Irving Freese, a Socialist, to the position of mayor. Similar examples could be cited from Bridgeport, Connecticut; Reading, Pennsylvania; and particularly, Milwaukee, Wisconsin. In all the cases mentioned the issue was one of good government, not of Socialist party doctrines and policies. Such has been the general strategy of farmer and labor parties. But the particular tactics used have not always furthered the general strategical objectives. We now turn our attention to a consideration of some of the tactics, with particular reference to the test of pragmatism—have the tactics worked or failed?

PURE AND SIMPLE UNIONISM

The age-old philosophy of the American Federation of Labor has done much to retard the creation of strong labor parties, based upon union membership. Despite persistent attempts by the Socialist Laborites, Socialists, and other groups to commit the AFL to political action, that organization has, with only one major exception, pursued steadily a policy of pure and simple unionism. The objective, that is to say, was felt to be working conditions, wages, and hours. As president of the AFL from its formation in 1886 until his death in 1924, with the exception of one year, Samuel Gompers guided his organization into non-partisan channels. The policy of the AFL as stated very simply by Gompers was "that economic organization and control over economic power were the fulcrum which made possible influence and power in all other fields. Control over the basic things of life gives power that may be used for good in every relationship of life." [3]

In 1924 the executive council of the Federation endorsed Senator LaFollette, but was careful to make clear that the endorsement was for the Senator and his running mate, only, and was not to be construed as an endorsement of the Socialist party. In the fall of 1947 the Federation voted at its convention in San Francisco to raise a "war chest" of one million dollars to combat the Taft-Hartley Act of 1947 through an organization known as Labor's League for Political Education. While this move was felt by many to be directed against the Republican party, it also was intended as an attack on anti-Administration Southern Democrats.[4] This action was, in fact, thoroughly in accord with traditional AFL tactics of rewarding friends and punishing enemies.

The Federation has, however, steadfastly refused to cooperate in the formation of any farmer-labor third party and has continued to operate on the basis of pressure group tactics within the framework of the two major parties. It has occasionally singled out individual Senators and Congressmen for defeat, and has cooperated with other unions and groups on a limited scale of political action.

The Political Action Committee of the Congress of Industrial Organization, on the other hand, has committed itself since birth in 1943 to aggressive political action. The PAC, like the AFL, subscribes to a non-partisan policy in that it refuses to endorse unconditionally any major party. Yet it operated in New York State as a partner of the American

[3] Quoted in Nathan Fine, *Labor and Farmer Parties in the United States, 1828–1928* (New York: Rand School of Social Science, 1928), p. 127.
[4] The AFL plan to try to defeat those who had voted for the Taft-Hartley Act was presented as a non-partisan move. In practice, however, the Republican party, particularly its leadership in Congress, bore the brunt of the attack. To this limited extent the AFL departed from its nonpartisan policy. But Southern Democrats were also under fire from the AFL.

Labor party until January of 1948, at which time the withdrawal of the Amalgamated Clothing Workers of America from the Labor party made it clear that the PAC would not support Henry A. Wallace as an independent Presidential candidate.

DISSENSION AND FACTIONALISM

When viewed historically, one of the outstanding characteristics of farmer and labor parties has been their lack of capacity for unity. Dissension and factionalism are as old as the movements themselves and show no signs of abating. It is debatable, however, whether doctrinal parties are more susceptible to fission than the usual major party. Fissions frequently occur within the major parties, even on a national scale.

The establishment in the 1948 campaign of two new third parties, each drawing primarily from the ranks of the Democrats, furnishes a somewhat extreme example. But the Liberal Republicans, the Silver Republicans, and the 1912 Progressives also owed their existence to deep party splits. The main difference between fission in a minor and in a major party seems to be this: Fission in a minor party usually results in wounds that cannot be healed; fission in a major party is usually eventually overcome by the strong centripetal forces inherent in both the manner of choosing a President and in the winner-takes-all system of state and Congressional elections. In contrast to a minor party, fission in the ranks of a major party may cost an election, but has not thus far since the Civil War been fatal. Several examples of factionalism inside farmer and labor parties are representative enough to be worth citing in detail. An early example is that of the Workingmen's party of New York more than a century ago.

In the election of 1829 the party was strong enough to elect the president of the Carpenters' Union, Ebenezer Ford, to the New York State assembly. This result naturally encouraged many party members to envisage further successes. Yet, two months after the election, in December, 1829, the party split when Thomas Skidmore, one of the leaders of the movement, left the organization to form the Poor Man's party. In 1830 a new split occurred, this time over details of the education program sponsored by Robert Dale Owen and Frances Wright. In the fall election of 1830 three Workingmen's tickets were in the field. The schisms contributed to the fact that the regular Democratic organization swept New York City, captured a majority of the seats in the state assembly, and won the governorship.[5]

A more recent example of factionalism is presented by the case of the American Labor party of New York State. Created in 1936 out of trade unions, some Old Guard Socialists, and Labor's Non-Partisan League (CIO), the American Labor party seemed destined for a while to be on the verge of demonstrating *bona fide* trade union solidarity. The largest and greatest financial contributor in the beginning was the International Ladies Garment Workers of America, and all the state officers were union officials.

In 1938, because of failure to poll 50,000 votes, the legal requirement, the Communists lost their place as a legal party in the State of New York. It was at once charged by the opponents of the ALP that the Communists were infiltrating into the party and were taking it over. In any event the issue of communism became heated and bitter. When the left wing of the ALP won 625 out of 750 seats to the state convention in the 1944 primaries, the right wing of the party formally seceded and in May organized the Liberal party. About a

[5] The labor political history of this period is to be found in Fine, *op. cit.*, and in John R. Commons and associates, *History of Labour in the United States* (4 vols., New York: Macmillan, 1918–1935), as well as in the standard histories of the period. Marxist interpretations are to be found in Philip S. Foner, *History of the Labor Movement in the United States* (New York: International Publishers, 1947, Vol. I), and in Alden Whitman, *Labor Parties, 1827–1834* (New York: International Publishers, 1943).

third of the delegates to the Liberal party convention came from AFL and CIO affiliated organizations and the remainder from various political and civic groups.

The differences between the ALP and the Liberal party were accentuated because of the struggle between two strong personalities. Sidney Hillman, head of the ACWA, and chairman of the CIO's Political Action Committee, was made state chairman of the American Labor party. At the same time David Dubinsky, the president of the International Ladies Garment Workers, accepted a position as first vice-chairman of the Liberal party.

The differences between the two rival parties were further increased in January of 1948 when the Amalgamated Clothing Workers left the ALP because they refused to go along with an ALP endorsement of Henry Wallace. The prospect of labor solidarity for political action thus disappeared, at least temporarily, from the scene in New York State.

FUSION

One of the greatest menaces to farmer and labor parties is the prospect of fusion with a major party. The classic example of fusion which brought with it disaster comes from the Populists. We are not here concerned with a repetition of the well-known facts of the election itself, but only with the results of fusion as illustrated by the fate of the People's party.

The problems raised by fusion were faced by the Populists virtually from the very beginning of their party. In the campaign of 1892 the Populists ran a former Union general, James B. Weaver, of Iowa, for President, and a former Confederate general, James G. Field, of Virginia, for Vice-President. It was the general policy of the party to array itself with the weaker of the two older parties in each locality. Thus, as a rule the Populists in the South cooperated with the Republicans; in the North, and especially in the new West, they at-

STRATEGY AND TACTICS 111

tempted either to fuse with the Democratic party or take it over.[6]

The issue of fusion on a national scale arose in 1896. The Populists held their national nominating convention *after* the Democrats had nominated a silver Democrat, William Jennings Bryan, of Nebraska. Silver, which had originally been considered by farmers' organizations an effective means of capturing votes, had by 1896 captured the People's party. When the Democrats nominated a silver man, they in effect stole the thunder of the Populists.

At the St. Louis convention held late in July, the majority of the delegates favored fusion with the Democrats, seeing in such fusion a fighting chance of sweeping the country for silver and inflation. But stern opposition came from the "middle-of-the-road" group, which opposed both fusion and cooperation. The main bloc of such delegates came from the South, where it was widely felt that fusion with the Democrats nationally would tend to destroy the Populist movement in the Southern states.

In an effort to show that the party had not surrendered to the Democrats but had merely made a marriage of convenience, the Southern Populists centered their efforts upon trying to nominate Thomas Watson of Georgia as Vice-Presidential candidate. In this they were successful, since the party refused to go along with Arthur Sewall of Maine, the Democratic Vice-Presidential nominee. An effort was made in several states to separate the Bryan-Sewall from the Bryan-Watson ballots in tallying electoral votes.

Cooperation between the Populists and Democrats in the North and the West was easy. It was a different situation in the South. The Populists were supposed to cooperate with the Republicans in the state elections and with the Democrats in the

[6] This account follows that given in Solon J. Buck, *The Agrarian Crusade* (New Haven: Yale University Press, 1920), pp. 149–50.

national election. But of all the Southern states and Border states it was possible for the Populists to agree on joint electoral tickets with the Democrats only in Arkansas, Kentucky, Louisiana, Missouri, and North Carolina.

The rest is summarized by Solon J. Buck: "The People's Party had staked all on a throw of the dice and had lost. It had given its life as a political organization to further the election of Bryan, and he had not been elected." [7]

The dire forebodings of the Southern Populists were also borne out. In 1896 the Populists, allied with the Republicans, swept North Carolina. Two years later, over the Negro issue, they were cast out of power. This served as a lesson for other Southern states and very shortly Southern Populism broke down.

The Populist party itself continued to lead a fitful existence until 1912, when the ghost was given up for good, and the old-timers could debate at their leisure the wisdom of fusion in '96.

ABSORPTION OF ISSUES

While the case of the Populists in 1896 is the outstanding example of the stealing of the thunder of a minor party by a major party, to the political detriment of the former, it is not necessary to go so far back in our history to find additional illustrations. The 1932 Presidential campaign furnishes some valuable material.

In accordance with the usually accepted cliché, it was to be expected that minor parties of the farmer-labor type would blossom forth during the darkest days of the depression. This did, indeed, happen in the spring and summer of 1932. There were 26 different third parties in the field during the summer and there were 21 different party names on the ballots of the various states at the time of the November election.

The Socialist party ran its perennial candidate, Norman

[7] Buck, *op. cit.*, p. 191.

STRATEGY AND TACTICS 113

Thomas; the Communists ran William Z. Foster; and the Socialist Labor party supported Verne L. Reynolds. A Farmer-Labor ticket (not related to the Minnesota party) was headed by Colonel Frank E. Webb and Jacob S. Coxey. When Webb was ousted, his place was taken by Coxey as Presidential candidate. The Farmer-Labor party presented the same ideas of agrarian "socialism" as did the Minnesota Farmer-Labor party.

Other farmer-labor parties took to the field. The Liberty party nominated the aged William "Coin" Harvey, a leading nineteenth century free silverite, at a convention featuring representation from 25 states. This party revived the old issue of free silver at the ratio of sixteen to one, called for inflation of paper currency, government ownership of banks and utilities, abolition of taxes, and a five-year moratorium on all private debts, including mortgages.

The discontented were offered another candidacy when the Jobless party nominated the Reverend James R. Cox, pastor of St. Patrick's Roman Catholic Church in Pittsburgh. Chief plank of the party's platform was a demand for generous relief.

The point which has been established is that there were a great many parties campaigning in the 1932 election for the votes of those suffering from unemployment and other forms of economic distress. Yet, as two authorities put it in their excellent joint study: "All of the minor parties were checked in their development by the almost instantaneous 'surrender' of the Democratic party to their demands." [8]

This "surrender" was very profitable to the Democratic party at the point where political profits are calculated—the polls. The Socialist party received 885,000 votes; the Communist, 103,000; the Socialist Labor party, 34,000; the Liberty party, 52,000; the Farmer-Labor party, 6,000; and the Jobless

[8] Roy V. Peel and Thomas C. Donnelly, *The 1932 Campaign* (New York: Richard R. Smith, 1935), p. 207.

party, 740. The Democratic candidate, Franklin Roosevelt, received a plurality of seven million votes over his Republican rival, Herbert Hoover, or 57.4 per cent of the total vote cast.

The significance of the 1932 election from the point of view of farmer-labor parties lies, then, in this outstanding fact: When actively challenged by parties of discontent at the bottom of a depression, one of the major parties swung sufficiently to the left to absorb the votes of the greatest number of the discontented. Thus parties which might have been expected to thrive on conditions of economic distress found their most vital issues taken over by one of the great parties.

This capacity of nearly unlimited flexibility by one or the other of the major parties illustrates an outstanding characteristic of our party system. This flexibility may be attributed to the fact that the principal objective of a major party is *to win an election.*

PROBLEMS OF ORGANIZATION

The legal and psychological difficulties which confront farmer-labor parties are discussed elsewhere. But it is appropriate at this point to call attention to some of the organizational difficulties which have been faced by such parties in their attempts to compete in elections.

It is an acceptable generality that most farmer-labor parties, excepting those of the ideological left, usually suffer from inferior organization. This is particularly the case with one-campaign parties competing in national elections. To cite a leading example, the LaFollette Progressives of 1924 did not come into being until the Cleveland convention of July 4.[9] This left less than four months available for the creation of a national organization, obviously an insufficient amount of

[9] A work by Kenneth C. MacKay, *The Progressive Movement of 1924* (New York: Columbia University Press, 1947), gives the history of the LaFollette campaign.

time. LaFollette had difficulty in complying with the various state election laws and was finally forced to run on a variety of labels—Progressive, Independent, Independent-Progressive, and Socialist. What saved the day for the Progressives was the fact that the Socialists, who supported LaFollette, has already established state machinery in forty-four states. That was not, of course, universally satisfactory to the Progressives, and probably cost them a certain number of votes.

Another difficulty, inherent in hasty organization, was the problem of raising funds. The national committees of the Republican and Democratic parties spent, respectively, $4,270,000, and $903,000. The Progressives ran a poor third with expenditures of $221,000.[10] The figures for the major parties are an understatement, because they include only expenditures of the national committees, omitting state and local expenditures to assist the national tickets.

Another major difficulty for LaFollette lay in the absence of adequate supporting tickets. The Progressives nominated LaFollette and Wheeler but they did not organize a new party. This meant that support from below was not dependable and that to vote Progressive usually included the necessity of splitting a ticket.

Without attempting to go into the subject thoroughly, it should also be noted that farmer and labor parties do not ordinarily enjoy a "good press." The only national newspaper chain which supported LaFollette was the Scripps-Howard organization. One student of the LaFollette movement has attributed the Progressives' success in carrying Cleveland in large measure to the vigorous support the Wisconsin Senator received from the Cleveland *Press*.[11]

The 1924 Progressives have not been alone in their belief that press coverage of their activities was inadequate and

[10] Mackay, *op. cit.*, p. 185. [11] *Ibid.*, p. 210.

generally unsympathetic. The charge was also made by the Populists, the parties of the left, and the Henry Wallace party of 1948.

A brief comparison between the Progressives of 1912 and the Progressives of 1924 illustrates the importance of a functioning organization before a national nomination is made. When Theodore Roosevelt seceded from the Republican party in 1912 in what was essentially a party schism, he took with him a goodly number of local bosses and some state machines. This meant that Roosevelt was assured of considerable support at the local and state levels from the very beginning of the campaign. In addition, he received very ample financial backing and the support of a great many well-known Republican newspapers. LaFollette, on the other hand, had to start almost from scratch.

The laying of previous groundwork is thus of fundamental importance if a farmer-labor party is to do well in a national election. In his analysis of the 1924 campaign, the distinguished scholar Arthur Macmahon had this to say of the relationship between advance activity in certain Eastern industrial areas, and dividends for the Progressives on election day in the same localities:

In nearly every one of these areas some influence had been at work in advance of the campaign, preparing at least the raw materials of organization, like the longstanding movements of social reform and labor leadership in Cleveland, the constant efforts at organization on the part of the Farmer-Labor Party in northern Illinois for over five years, the unusually strong interest of the Pennsylvania State Federation of Labor, under Mr. Maurer, in politics and in Rochester, the movement which at the same election returned Mr. Jacobstein to Congress with combined Democratic and Socialist endorsement. The spotty character of the LaFollette vote in the East was not accidental.[12]

[12] Arthur Macmahon, "Political Parties and Elections," Special Supplement to *Political Science Quarterly*, XL (1925), 51. Reprinted by permission of the publishers.

It is interesting to observe that the groundwork for the Henry A. Wallace campaign was laid by the Progressive Citizens of America, a nation-wide organization with considerable previous experience in local, state, and Congressional elections. The refusal of the CIO's Political Action Committee to endorse Wallace eliminated the possibility of still wider support based on trade union membership.

BALANCE OF POWER TACTICS

The three ways in which any political party may achieve control are (1) to win a national election; (2) to win regional or state supremacy; (3) to hold the balance of power in either national or state elections.

There is no instance in which a farmer or labor party succeeded in winning a national election, although the Populist-Democratic coalition of 1896 came reasonably close to achieving victory. There are several instances in which Granger or Independent parties, and later, the People's party, controlled whole states either through the plurality (winner-takes-all) principle or through holding in effect the balance of power as between the major parties.

Recent party history affords two and—depending upon the mode of calculation—possibly three examples of domination or partial domination of a state by a farmer-labor party. These are the reigns of the Minnesota Farmer-Labor party, from 1922 to 1939, especially from 1931 to 1939; the Wisconsin Progressive party, from 1934 to 1938; and the Non-Partisan League of North Dakota, from 1916 until the present time (through capturing the Republican party).

There is some evidence at present which points to an emerging pattern for future activity by farmer or labor parties through the balance-of-power technique. The best available example is the recent record of the American Labor party of New York State.

Since the electoral laws of New York permit a minor party to nominate the candidate of a major party, it has been possible for the ALP to prove by the election returns on several occasions that it has held the balance of power. In the 1936 gubernatorial election the Democrats could have won without ALP support, unless all ALP voters had supported the Republicans. The returns (for major candidates only) were:

Lehman, Democrat	2,708,383
Lehman, ALP	263,092
Bleakley, Republican	2,450,104

But two years later, in the gubernatorial election of 1938, the ALP very clearly held the balance of power. By its endorsement of Herbert Lehman, the Democratic candidate, the ALP swung 419,979 votes into the Lehman camp. This amounted to 8.9 per cent of the total vote for the two major candidates and was sufficient to defeat the Republican candidate, Thomas E. Dewey. The returns were as follows:

Lehman, Democrat	1,971,307
Lehman, ALP	419,979
Dewey, Republican	2,302,505
Dewey, Progressive	24,387

The ALP has enjoyed a similarly impressive record in Presidential races. In the 1940 election the leading candidates received the following numbers of votes:

Roosevelt, Democrat	2,834,500
Roosevelt, ALP	417,418
Willkie, Republican	3,027,478

Thus, the ALP vote was sufficient by a considerable margin to carry New York State for Roosevelt.

The same technique was successfully utilized in the 1944 Presidential contest in New York. In this election the newly formed Liberal party joined the ALP in endorsing the Demo-

STRATEGY AND TACTICS

cratic nominee, Franklin D. Roosevelt. The returns were as follows:

Roosevelt, Democrat	2,478,598
Roosevelt, ALP	496,405
Roosevelt, Liberal	329,235
Dewey, Republican	2,987,647

Roosevelt received 39.4 per cent of the total vote cast for major candidates as a Democrat; 7.9 per cent as an American Laborite; and 5.2 per cent as a Liberal. This gave him a total of 52.5 per cent and consequently the 47 electoral votes of New York State.

The ALP supported the Progressive candidate, Henry A. Wallace, while the Liberal party endorsed President Harry S. Truman, in the 1948 contest. The ALP polled 509,559 votes to 222,562 for the Liberals. Since Governor Dewey carried New York State by only about 60,000 votes, it is clear that the ALP Wallace vote was sufficient to throw the 47 electoral votes of that state into the Republican column.

The voting strength of both the American Labor party and of the Liberal party is concentrated in New York City. These parties have therefore played very important roles in local elections. In the 1937 mayoralty election the ALP contributed 482,790 votes to LaGuardia, who was the Republican-ALP-Fusion candidate. This guaranteed LaGuardia's victory. In 1941, with a somewhat smaller vote, the ALP also clearly held the balance of power. The vote was as follows:

O'Dwyer, Democrat	1,054,235
LaGuardia, Republican	668,485
LaGuardia, ALP	435,374
LaGuardia, Fusion	63,367
LaGuardia, United City	19,292
LaGuardia,Total	1,186,518
LaGuardia majority over O'Dwyer	132,283

When the American Labor party ran its own candidate for governor in 1942, the Republican candidate, Thomas E.

Dewey, triumphed. But the total of the Democratic and ALP votes would not have been sufficient to offset the Republican lead.

In the 1945 mayoralty election in New York City, which featured a warmly contested battle among three candidates, the ALP endorsed O'Dwyer, the Democratic candidate, and gave him 257,930 votes. The Liberal party, which supported Goldstein, the Republican candidate, amassed the total of 122,316 votes. The O'Dwyer plurality in the election was 693,-717, which the New York *Times* hailed as a "decided victory for the PAC and the American Labor party." [13]

During the 1946 gubernatorial election the ALP and the Liberal party both endorsed the Democratic candidate, Mead, against the Republican incumbent, Governor Dewey. The state-wide ALP vote was about 425,000 and the state-wide Liberal party vote reached 180,000. Nevertheless, the popularity of Governor Dewey was so great that he carried the state by a majority of 656,000 votes over his Democratic-ALP-Liberal rival.

The most impressive local showing of the ALP in the 1948 New York City elections occurred in the contest for the Surrogate of New York County (Manhattan). In an unusually close battle, the 97,000 votes cast for the ALP candidate resulted in a Republican victory for the first time in half a century. The 63,000 votes cast for the victorious Republican on the Liberal party line contributed the margin of victory. In the Congressional election, the ALP reelected Vito Marcantonio as representative from the 18th Congressional district. However, in most New York City contests the ALP lost the balance of power position it had hitherto enjoyed.

What the future will bring in New York State, in view of the rivalry between the American Labor and the Liberal parties, cannot be foreseen. But it is quite clear that in 1937,

[13] New York *Times*, Nov. 7, 1945.

STRATEGY AND TACTICS

1941, and to a lesser extent in 1945, the ALP held the margin of victory in the New York City mayoralty contests; it gave Lehman sufficient votes to defeat Dewey in the gubernatorial race in 1938; and it made handsome contributions to the Roosevelt Presidential victories in 1940 and 1944.

It is interesting to speculate on the membership of the ALP and the Liberal party. The available evidence points to the conclusion that the membership of these groups has not, primarily, been drawn from the ranks of the Democratic party. This is so even though both parties gave electoral support to President Roosevelt in 1944, and though the ALP gave its support to the same man in 1940.

The ALP showed in the 1948 Presidential campaign that it did not consider itself merely an adjunct of the Democratic party when it supported the candidacy of Henry A. Wallace. In the 1945 New York City mayoralty contest the Liberal party supported the Republican candidate. Earlier, the ALP had supported LaGuardia as the Republican-ALP nominee for mayor. It must always be remembered that in New York City there has been for generations a tendency for reform groups of all complexions to unite, usually under the Republican aegis, in the political struggle against Tammany.

In New York State in the largest cities, there has always been a sizable urban protest vote for parties of the left, whether Socialist, Communist, or some other group. The tradition of leftist politics in New York has been ephemeral but continuous. Since 1936, the vote for the Socialists and the Communists in New York State has all but disappeared. In 1932, for instance, the Socialists polled 177,000 votes, a substantial increase over their total in the preceding Presidential election. In 1936, the first Presidential election in which the ALP appeared on the ballot, the vote for the Socialists dropped drastically to 87,000, and by 1944 it was reduced to 10,600. The Communist party disappeared from the ballot in 1940.

The causes for these decreases were undoubtedly many and varied; yet it seems reasonable to assume that much of the discontent formerly channeled by these parties tended subseqently to operate through the ALP and Liberal parties.

Trade unionism, particularly in New York City, has ordinarily sought a political outlet. Thus the ideology of some of the unions which at various times have supported either the ALP or the Liberal party—particularly the International Ladies Garment Workers and the Amalgamated Clothing Workers—is much more in tune with the philosophy of social democracy than with the politics of Tammany Hall and the Democratic party. A change in policy by the national or state Democratic organizations, whereby they ceased to push desirable labor legislation in Washington or Albany, would quickly disaffect the labor groups.

For these reasons, it seems clear that to consider the supporters of either the ALP or the Liberal party generally to be Democrats with another name is in error. On the contrary, the available evidence points to the conclusion that when united in a single endorsement the ALP and Liberal parties may under favorable conditions be able to control New York State politics, sometimes by giving, sometimes by withholding an endorsement.

The balance-of-power technique which has been put to great use in New York State may also be in process of achievement in Illinois. While it is still too early to venture predictions, it appears that the Progressive party of Illinois, which polled 313,000 votes for its top candidate in the Cook County judicial elections of November, 1947, may attempt the same strategy.

In the 1948 Presidential and Senatorial elections the Progressive candidates were not on the ballot in Illinois. Since the Democratic plurality in the Presidential race was only 34,000, it appears likely that the Progressives, had they been on the ballot, would have cost President Truman the 28 electoral votes of Illinois.

The Farmer-Labor elements of the merged (in 1944) Minnesota Democratic-Farmer-Labor party may likewise aspire to a controlling position. In unpredictable California there are signs that the Independent Progressive party (set up to support Henry A. Wallace) may continue on the same basis. Wallace received 190,000 votes in California, but he did not hold the balance of power. President Truman carried the state by 18,000 votes.

CONCLUSIONS

From the preceding analysis several conclusions warrant emphasis. A farmer or labor party which is really out to win an election, rather than merely to agitate, tends to seek the broadest possible basis of support. This is made difficult by a number of considerations, of which two of the most important are factionalism and fusion. Factionalism is almost certain to result in defeat, and fusion is almost certain to result in absorption. Between this Scylla and Charybdis many a farmer or labor movement has come to disaster; few have been able to steer an even course through the channel.

The problem is made extremely difficult because of the tendency of the major parties to capitalize on any really popular issue. This is to say, again, that the flexibility of the major parties is one of the chief characteristics of the American party system.

In addition to the obstacles which are created by the presence of the two-party system itself, the problems of organization and finance are almost insolvable. Few people are willing to spend time and money on a losing organization.

Despite these formidable difficulties, a new pattern of farmer and labor party politics may be emerging. This is the balance-of-power technique. The idea is by no means new, but its successful practice in New York State leads to the belief that it may spread. It may also be successful in other states which have large metropolitan areas and great numbers of unionized workers.

In the past the successful application of the balance-of-power principle has been largely confined to rural areas. It seems likely that some workers' parties in large cities will now assay the same role. If American politics continues its present trend of becoming more urban and less rural, with increasing emphasis on class as against sectional issues, balance-of-power techniques in all probability will be exploited on an increasingly larger basis. The potential limit to the application of such techniques is far greater than the successful application to date.

7

LEGAL BARRIERS

ONE WAY in which politicians of major parties can eliminate or curtail competition is to make it difficult for new parties to appear on the ballot. There is some evidence that in many states it is now more difficult than formerly for a minority party candidate or an independent candidate to appear on the ballot. The restrictions which have arisen may and do frequently hamper small farmer and labor parties. But groups with mass support—such as the American Labor Party of New York—find themselves but little affected by the legal restrictions.

A political party may be placed on an election ballot in two ways. The first method is by polling a certain percentage or a definite number of votes cast at a preceding election. The second method is by presenting, to the appropriate officials, petitions with the prescribed number of signatures.[1] The

[1] Much of the data in this section has been taken from a pamphlet, "Minority Parties on the Ballot," published by the American Civil Liberties Union (New York: August, 1940; revised, April, 1941; revised, January, 1943). Other material has been gathered from current periodicals, newspapers, and literature distributed by the Wallace party headquarters in New York City.

The New Republic in the spring and summer of 1948 devoted much attention to the difficulties encountered by the new Progressive party in getting on the ballot. In Ohio the legal barriers were particularly difficult to overcome. A ruling of the Ohio Supreme Court was required to reverse a decision of the Secretary of State. Secretary of State Hummel based his decision on two grounds, first, that the party candidates had not been nominated at conventions by delegates chosen by voters in the primary elections, and secondly, that some of the Wallace sponsors could not qualify under the laws of Ohio because they were Communists. On these points, see "New Party in Ohio," *New Republic*, CXIX, No. 6 (Aug. 9, 1948), 10; also, Section 4785–100a, *Ohio Election Laws Annotated* (1943). This section of the law relates to parties advocating overthrow of the government by force or violence.

The view that small parties may be on the wane is ably presented by Hugh Bone in "Small Political Parties Casualties of War?" *National Municipal Review*, XXXII, No. 10 (November, 1943), 524–528, 565.

analysis which follows is intended to show the types of legal obstacles frequently placed in the path of minority parties. Since the election laws of the various states are often changed, the following examples are not intended to constitute a complete listing but rather representative illustrations. The restrictions cited were in effect as of January, 1948.

In some states the number of petitioners needed to place a party on the ballot has reached a very high figure. In other states there is a provision that no petitioners shall have registered in a primary of any other party. Other difficulties arise. Courts have frequently ruled that signatures on certain petitions were obtained through fraud or misrepresentation. On occasion petitions have been disqualified because of the requirement that all circulators of petitions must swear that they knew all the signers to be qualified voters living in a particular place. In yet other cases the filing fees have been so high as to discourage minor party candidacies.

Publicity has sometimes been employed in an effort to embarrass signers of petitions. An example of this occurred in the spring of 1948, when the Pittsburgh *Press*, a Scripps-Howard newspaper, published the names and addresses of persons who had signed a petition circulated on behalf of Henry A. Wallace. The purpose was obvious.

Where a percentage of votes cast at a preceding election is a requisite for appearing on the ballot, the percentage required varies from one to twenty. In terms of the sheer number of signatures required, the requirement of California when translated from a percentage into figures is impressive. About 270,000 signatures are required to put a new party on the ballot. To put a new party on the ballot in New York 12,000 signatures are required on a petition, including fifty from each county. Yet, for a party to maintain its place on the ballot, it must poll not less than 50,000 votes.

In Illinois the requirement calls for 25,000 signatures on a nominating petition for state-wide office. This requirement appears to be easily met, until it is noted that the law requires not less than two hundred signers from each of fifty counties. Such a condition tends to make it difficult for a purely farmer or a purely labor party to get on the ballot. The 1948 Progressives failed to qualify in Illinois. In Rhode Island a petitioner for a new party must not have been affiliated with another party for a preceding period of twenty-six months.

About half the states require filing fees from candidates. The original and still chief purpose of such a requirement is to keep the ballot from being cluttered with the names of candidates who are not really trying to win the election. As a consequence, the filing fee in most states is nominal. In some states the fee is fixed by a percentage of the salary of the office sought. This is usually from one to three per cent. But a candidate running for United States Senator from Tennessee must pay as much as $500 at the time of filing.

Additional restrictions were imposed during World War II on parties alleged to be extremist. Even before 1939 several states, including Arkansas, Delaware, Indiana, and Tennessee, excluded by law candidates of parties which advocated the "overthrow of government by force and violence." This provision was aimed principally at the Communist party. After 1940 a considerable number of other states followed suit, including by June, 1941, the states of Arizona, Georgia, Kentucky, and New Mexico (all by administrative rulings) and Illinois, Kansas, Ohio, Oregon, Oklahoma, Texas, Wisconsin, and Wyoming (by law). A similar California law was invalidated by the California Supreme Court in July, 1942, on the ground that it was discriminatory. A United States District Court invalidated a parallel law in Illinois in the fall of 1942. By January, 1949, fourteen states had laws on their books which operated to exclude the Communist party, among others, from the ballot.[2]

---

[2] The states whose laws as of January, 1949, excluded from the ballot political parties and candidates for public office advocating doctrines inimical or

The principal objection to restrictive legislation centers on a denial of freedom of political expression. It is also urged that the laws work a hardship on independent nominees and on all new parties.

As a solution to some of the difficulties, the American Civil Liberties Union has proposed the passage of a model election law containing provisions for independent nominations and for new parties.[3] This model law, which was based on the Wisconsin statute, proposed that nominating petitions for state-wide office require one-tenth of one per cent of the total number of votes cast in the last general election for governor, with a minimum of 1,500 signatures and a provision that not more than 10,000 signatures be required. The proposal was also presented that nominating papers should be filed for candidates to be voted on throughout the state not more than forty nor less than thirty-two days before the election.

The American Civil Liberties Union also called attention to some additional requirements which it believed to be objectionable. Arkansas requires a fee from Presidential electors, which amounts to $337.50 for a full slate of nine electors. Indiana requires only a small number of signatures (8,000 to 10,000) on a nominating petition, but since each signature must be individually notarized, the expense is considerable. In Michigan a new party must file its petition six months before election day, which is long before the major parties have made their own nominations. The Mississippi statute requiring fifty signatures for an independent nomination has never been constructed by the courts, so that it is not certain whether

---

hostile to the American form of government were Arkansas, California, Delaware, Illinois, Indiana, Kansas, Ohio, Oklahoma, Oregon, Pennsylvania, Tennessee, Texas, Wisconsin, and Wyoming. This list is taken from the report of the Maryland Commission on Subversive Activities, January, 1949, as summarized by E. T. Baker, "Maryland Betrays Its Past," *New Republic*, CXX, No. 17 (April 25, 1949), 16.

[3] The model law is reprinted in the pamphlet, "Minority Parties on the Ballot."

the electoral law means fifty signatures from the whole state or from each of the eighty-two counties of the state.

Among the most liberal of all states would appear to be South Carolina. All a candidate need do is to print ballots and distribute them to the election precincts. But since the voter must ask for the ballot of the party whose candidates he wishes to support, he does not have a secret ballot and he finds it impossible to split his vote. South Carolina is the only state in the Union which does not observe the principles of the Australian ballot.

There is one additional method by which a minor party may secure a place on the ballot. Most states permit the use of the write-in vote, so that any minority party candidate who files nomination papers under the name of his party and secures the requisite number or percentage of votes through write-ins will find his party automatically on the ballot at the next election. While this procedure is technically possible, there is no example of a party's having achieved a place on the ballot in this manner. This means that the practical method of getting on the ballot is to use the device of a nominating petition.

Except that a state may not prohibit an otherwise duly qualified person from voting because of race, color, previous condition of servitude, or sex, a state may establish qualifications for voting. It is sometimes charged that various suffrage requirements penalize farmer and labor parties by preventing potential supporters from voting. The argument falls into two divisions: The first concerns limitations on the voting of Negroes and a considerable number of poor whites in the South, while the second deals with residential requirements that prevent transients from voting.

In examining the first part of the argument it is advisable to review briefly the nature of the one-party system in the South. According to Paul Lewinson in his excellent study, *Race,*

Class, and Party, the South enjoyed a brief period of two-party politics after the Civil War and prior to the Reconstruction Acts.[4] The two-party system was similar to that which had existed prior to 1860. But the Reconstruction Acts, intimidating pressure of the Union Army of Occupation, and large scale voting by Negroes soon united the whites into a single party. The freedmen were, of course, encouraged and instructed by Union troops to support the Republicans. Thus the system of having two white parties broke down completely.

When President Hayes in 1877 withdrew the last of the Union troops, the final impediment to white supremacy vanished. By this time it is important to note that Negro voting was no longer able single-handedly to control the election results in a single Southern state. Two-party politics in which white voters competed with other white voters returned to the Southern scene. The struggle was between the Granger and later the Populist groups, on the one hand, and the so-called "Bourbon Democracy" of the established property groups, on the other.

One result of this political realignment was the possibility of Negro voters holding the balance of power between the parties of competing whites. This possibility was widely recognized as a potential challenge to white supremacy.

In the 1890 elections the Southern Alliance men took control of the legislatures of eight Southern states, elected three governors, and helped elect a fourth. In South Carolina the Tillman Democrats scored the most notable Alliance successes, but they did so while maintaining a strict anti-Negro attitude. In the Presidential election of 1892 the battle between the Southern Democrats and the Southern Populists was heated, bitter, and deadly serious. Appeals by the Popu-

[4] Paul Lewinson, *Race, Class and Party* (New York: Oxford University Press, 1932). See also V. O. Key, Jr., *Southern Politics* (New York: Knopf, 1949).

lists for support at the polls by members of the Colored Alliance, an adjunct of the Southern Alliance, only made matters worse. The election results showed that Democratic supremacy had been seriously threatened in both Georgia and Alabama, although the Populists failed to carry either state.

One of the outstanding electoral successes of the Populists occurred in the North Carolina state and Congressional elections of 1896. A Populist-Republican coalition swept the state, taking over the offices of governor and lieutenant governor, electing seven of nine Congressmen, and securing a majority in both houses of the state legislature.

Two years later—in 1898—the race question had become the chief issue in North Carolina. By capitalizing on this issue the Democrats returned to power throughout the state. They were able to return to power because the reunited whites largely supported the white supremacy Democratic ticket. The alliance of poor whites and Negroes had collapsed under the pressures generated by raising the race question.

It is in the light of the challenges to Bourbon Democracy by the Granger and Populist movements that poll taxes should be considered. The first post-Civil War poll tax was set up in Florida in 1889 and by 1901 eight Southern states used the poll tax payments as a requisite for voting. The poll tax in Georgia was abolished in 1945 under the vigorous leadership of Governor Ellis Arnall.

A significant feature of the poll tax is that it disfranchises large numbers of the poorer whites. One authority on the poll tax, Barry Bingham of the Louisville *Courier-Journal*, has estimated that "64 per cent of the white adult voters have been disfranchised in the poll tax states, and in every one of those states more whites than Negroes are barred from the ballot box as a direct result of this tax." [5] Thus, the effect of

[5] Barry Bingham, "Americans without Votes," in *The Poll Tax* (Washington: American Council on Public Affairs), p. 5.

poll taxes was not originally and is not now exclusively to prevent the Negro from voting.

The incidence of the poll tax on the voting for farmer and labor parties is not easily ascertainable. Voting statistics in Presidential elections would seem to indicate that there was frequently little change in the share of the vote going to farmer-labor parties after the tax was passed. Two notable exceptions occurred in Alabama and Mississippi.

Alabama adopted the poll tax in 1901, several years after the peak of Populist voting strength had passed in that state. Since that year, in Presidential elections, Alabama has never again given a significant share of its vote to a farmer-labor party. The passage of a poll tax by Mississippi in 1890 did not prevent a large vote for the Populists in 1892, but since that year there has been little support for parties of this type in Mississippi.

In Arkansas, however, the share of the total vote going to farmer-labor parties was higher in 1912 and in 1924 than it was in 1892, the year the poll tax was passed. Nor did the tax appear to deter would-be voters from supporting farmer-labor parties in Georgia. In spite of a poll tax in effect since before the Civil War, 20 per cent of Georgia's voters were able to cast their ballots for the Populists in 1892. In 1904 and 1908, some 17 and 13 per cent of the vote was polled by farmer-labor parties, a figure higher than that achieved in many of the Northern industrial states. On the other hand, in South Carolina, voting for farmer-labor candidates in Presidential elections was as insignificant before as it was after the passage of the poll tax.

Since the repeal of the poll tax in Georgia in 1945, there has not thus far been any relative increase in the voting for candidates of farmer and labor parties.

Of the 11 states mentioned, four have repealed the poll tax requirement as a prerequisite to voting. Repeal occurred

LEGAL BARRIERS

in North Carolina in 1920, in Louisiana in 1934, in Florida in 1937, and in Georgia in 1945. The Southern states which do not have poll taxes as a prerequisite to voting have a higher percentage of voter participation in Presidential elections than do the seven states which still maintain the tax. Yet the average participation in these relatively favored states is far less than that in other parts of the country. Thus, in 1944 the percentage of total population actually voting in North Carolina was 37; in Florida, 23; in Louisiana, 15. For the same election the percentages in other representative states were as follows: California, 43; Illinois, 54; New York, 51; South Dakota, 44.

These figures contain certain non-comparable factors, since they deal with voting participation in Presidential elections. White participation in the Democratic primaries of Mississippi and South Carolina is at a relatively high level, probably higher than that in many non-Southern states. Yet Mississippi has a poll tax while South Carolina levies no tax for voting in the primary.

It is nevertheless the case that many persons, white as well as Negro, cannot in practice participate in many Southern elections because of their failure to meet the poll tax requirement. But factors other than the poll tax are probably primarily responsible for the general low level of voter participation in the states of the South. The single-party system which normally prevails is perhaps one of the chief reasons. Since the causation of low voter-participation is multiple, it is difficult to isolate the poll tax and determine its exact effect upon voting today. It seems reasonable to suppose, however, that the repeal of various restrictive devices which limit the suffrage of poor whites and Negroes, including among them the poll taxes, would eventually raise the rate of participation in primary as well as in general elections throughout the South.

There is, however, no conclusive evidence to suppose that such measures would necessarily result in an appreciably higher percentage of votes being cast for candidates of farmer and labor parties. It appears more likely that the South would return to a two-party political system resembling that in most other sections of the country. There can be no question that the Negroes would especially benefit under such a system, but there is considerable doubt whether farmer and labor party candidates would benefit disproportionately.

The problem of disfranchisement is by no means limited to the South. At all times, but especially during depressions, transients have found it extremely difficult to vote. At an interstate conference held at Harrisburg, Pennsylvania, in August of 1932, it was estimated that more than one million migrants would not be able to meet the residential requirements for voting in the forthcoming Presidential election.

It may be argued with some plausibility that many of the migrants, unemployed and bitter, might have supported farmer or labor parties if they had been able to vote. Deliberate attempts, however, to disfranchise persons on relief by invoking ancient poor laws have usually been repelled. The most widely publicized such attempt occurred in 1932 when several Maine cities tried to invoke a law of 1830 which took paupers off the voting list. Most of the so-called paupers were simply unemployed workers temporarily on relief. Fortunately, the attempt to disfranchise this group was successfully challenged at the time.

It nevertheless remains true, particularly in depressed periods, that farmer and labor parties, and probably all protest parties, lost potential strength in the form of unemployed migrants because of strict residential requirements. The extreme length in residential requirements is found in some Southern states and in Rhode Island. In Rhode Island the law requires two years of residence before the non-property owner may be qualified to vote.

An unusual type of difficulty arose in Minnesota out of the peculiar law of that state governing election to the state legislature. The effect of the law on the Minnesota Farmer-Labor party was considerable.

The story goes back to 1920, when an independent group which accepted the principles of the Non-Partisan League drew up a full slate of candidates headed by Dr. Henrik Shipstead. This slate was entered in the Republican primaries but failed to receive party endorsement. The group then filed as independents in the regular election.

After the Republicans had won the election, by a narrow margin, they passed a law requiring any group to decide before a primary whether it was part of an existing party or was in fact an independent organization. The main point to the law was that a group could not be both.

One result of this law was that the Shipstead group organized the Farmer-Labor party. This party enjoyed considerable success from 1922 to 1939 in state politics. From 1931 to 1939 the Farmer-Labor party was ascendant in Minnesota but it never controlled the state legislature. In the 1931 and 1935 sessions of the legislature the party controlled neither house nor senate, and it controlled only the house in the 1933 and 1937 sessions. Meanwhile the party was electing and reelecting its candidates by great pluralities and majorities to state administrative offices.

This seeming paradox was attributed by Arthur Naftalin to four basic factors.[6] First, Minnesota during the entire period was a three-party state; secondly, the legislature had not reapportioned the state since 1913 (this in effect dis-

[6] Arthur Naftalin, "The Failure of the Farmer-Labor Party to Capture Control of the Minnesota Legislature," *American Political Science Review,* XXXVIII (February, 1944), 71–78.

criminated against the cities, where the Farmer-Labor party had its greatest strength); thirdly, the entire state senate in Minnesota is chosen for a four-year term; fourthly, Minnesota is one of the two states of the Union which choose a legislature on a non-partisan basis.

Naftalin found the key to the riddle in the last observation. The non-partisan elections to the state legislature resulted in an absence of party principle and party responsibility. This penalized an organization such as the Farmer-Labor party, which was inclined to adopt a strong position on controversial subjects. "Thus," says Naftalin, "legislative contests frequently are reduced to nothing more than popularity contests."[7]

SUMMARY

Various types of legal obstacles make it difficult for small parties of all kinds to appear on the ballot. When farmer and labor parties are weakly organized and have no great following, they often find it difficult to place their slates before the electorate. Similar difficulties arise with respect to independent candidates.

Nevertheless, a farmer or labor party which has a large following finds the legal obstacles irritating but not prohibitive. LaFollette slates appeared on the ballots of all the states except Louisiana, where it was necessary to write in the names of LaFollette electors. Similarly, the Wallace movement of 1948, with local groups working vigorously, was able, despite legal obstacles, to get on the ballot in all states except Illinois, Nebraska, and Oklahoma. The lesson for minor parties is that they can be sure of getting on the ballot only if they enjoy widespread support.

[7] Naftalin, *op. cit.*, p. 76. A brief discussion of the Minnesota Farmer-Labor party may be found in an article by Clarence A. Berdahl, "Party Membership in the United States," *American Political Science Review*, XXXVI (April, 1942), 241–262, especially 241–242.

8

THE COMPONENTS OF PROTEST VOTING: ECONOMIC DISCONTENT

AT THE POLLS the American public votes for the candidates or the parties which most closely represent the feelings and inclinations of each individual voter. In an indirect way, through the process of selecting their representatives, the public is voting upon the issues of the day.

Because of the secret ballot much of the interpretation of who votes why is reduced to speculation. To throw additional light on the various groups in the electorate to whom the type of economic and political changes often advocated by farmer and labor parties might be appealing, this chapter investigates pertinent questions asked in public opinion polls. Through public opinion polling, the issues are directly presented to a sample of the public. This sample, stratified to include representatives of our total population by age, sex, economic status, occupation, and geographic location in proportion to their numbers in the whole, is thus virtually a small replica of the national population, or in election surveys, the voting public. Representing as it does a statistical miniature of the public, it can then bebroken down and its component parts may be investigated.

All questions and answers printed here are taken from the *Fortune Survey*, unless otherwise indicated. These selected questions are by no means intended to be a definitive study of public opinion on the complicated and multifold questions of our national economic and political systems. Rather, a few questions out of many have been chosen to give an indication as to how farmers and workingmen, in contrast to one another and to other groups comprising the general public, react to the selected issues.

Among which segments of our population will a program calling for economic changes attract the greatest support? The following group of questions investigates attitudes toward four aspects of government intervention into economic affairs. They are all issues which have been repeatedly emphasized in the speeches and platforms of many of the minor parties, especially those which have attempted to attract the votes of farmers and workers.

Many of the questions were asked in 1939, the seventh year of the longest business depression to date. This year was one of great economic discontent, although the bottom of the depression occurred in 1932. Industry was slowly climbing back from a recession which followed what had seemed to be the start of a recovery. In 1939, wholesale prices were still low, especially those of farm products. While they never again reached the depths established in 1932, they were well below the levels of the 1920's. A more important factor making for economic discontent was the extent of unemployment. By 1940, more than seven million people were still seeking work or were employed on emergency public works programs. The times were the more depressing because seven years of the New Deal had failed to bring permanent recovery.

The following questions concern measures which both federal and state governments had to a large extent already put into effect. During the early days of the New Deal, unprecedented steps had been taken to alleviate the hardships of the unemployed. Public utility rates had been regulated by state governments for many years. The income tax, adopted in 1913 after an amendment to the Constitution, represented one form of wealth redistribution. Receipts from income taxes were increasingly used by Franklin D. Roosevelt's administration to finance social reform.

The American public in 1939 has this to say about various economic proposals:

June 1939*—People feel differently about how far a government should go. Here is a list of things some people believe in, others don't. Let's take them one at a time. Do you think our government should or should not—

	Should	Should not	Don't know or depends
Be responsible for seeing to it that everyone who wants to work has a job	61%	32%	7%
Regulate all public utility rates like electricity, gas, etc.	49	39	12
Redistribute wealth by heavy taxes on the rich	35	54	11

* Questions are dated according to their publication date in *Fortune*.

The issues presented by the pollers to the public, of course, did not suggest pioneer areas for governmental activity. Yet, in response to only one question, that concerning unemployment, did more than a majority of those interviewed favor extension of governmental activity.

Going far beyond mere government regulation, several farmer and labor parties in the twentieth century have advocated government ownership of the "means of production." Agitation in the nineteenth century was principally directed toward the correction of what were considered injustices and inequities under the existing capitalist structure. It remained for the Socialist party and Marxist splinter groups to rationalize these economic demands into a comprehensive system of government control of enterprise. Even such a widespread and popular party as the LaFollette Progressives of 1924 considered some form of nationalization of industry an important part of its over-all program. But this program has not found favor with the American public:

June 1939—On the following things we have found that people differ as to the *degree* to which government should function. Do you think the government should own and operate all, some, or none of—

	All	Some	None	Don't know or depends
The railroads	22%	12%	52%	14%
The insurance companies	13	14	62	11
The factories producing the essentials of life such as clothes, food	7	14	71	8

Here, without exception, majorities were against having the government take over fields traditionally left to private enterprise. The least opposition was felt toward government operation of railroads, probably because this public service industry has long been regulated by government. But some two-thirds of the people were clearly against the nationalization of manufacturing and insurance.

Different groups within the public, however, did not divide as consistently on these various measures. Table 2, giving a breakdown by economic status, shows that there were substantial differences between the poorest groups and the most prosperous. The gap is particularly noticeable on the subject of unemployment. The prosperous group, consisting primarily of the financially successful merchants, bankers, lawyers, doctors, farmers and their families, are able to afford most of the luxuries common to their community. They are shown to have been out of step with the general public on the issue of government responsibility for alleviating unemployment. The poor group, comprising those people who are able to afford the necessities of life only so long as their not-too-secure jobs last, felt understandingly vehement on this subject.

It is interesting to note that the lower middle class divided itself in almost the same ratio as did the public as a whole in answer to these questions. The lower middle class, as defined

TABLE 2

ECONOMIC CLASS AND OPINION ON THE ROLE OF GOVERNMENT
IN ECONOMIC AFFAIRS *
(June, 1939)

	Pros-perous	Upper Middle	Lower Middle	Poor	Nat'l Total
THE GOVERNMENT SHOULD:					
Guarantee employment					
Yes	39%	50%	60%	71%	61%
No	55	44	32	23	32
Don't know	6	6	8	6	7
Regulate utilities					
Yes	36	45	50	50	49
No	58	45	39	33	39
Don't know	6	10	11	17	12
Redistribute wealth					
Yes	17	28	34	46	35
No	76	64	56	40	54
Don't know	7	8	10	14	11
THE GOVERNMENT SHOULD OWN AND OPERATE:					
The factories					
All	1	3	7	10	7
Some	8	11	14	16	14
None	89	81	72	62	71
Don't know	2	5	7	12	8
Insurance companies					
All	5	9	13	18	13
Some	11	13	14	15	14
None	81	69	63	53	62
Don't know	3	9	10	14	11
The railroads					
All	13	16	22	27	22
Some	10	10	11	14	12
None	72	63	55	43	52
Don't know	5	11	12	16	14

* See pages 139 and 140 for full text of questions.

here, consists of those who earn enough and are secure enough to take most of the comforts and necessities of life for granted, but who have to struggle to obtain some of the simpler luxuries. It includes such people as minor white collar workers, proprietors of small local stores, skilled factory workers, and their families. This group might be called the "great middle class."

Within ten years, the economic situation changed greatly. The United States emerged from the war to enter a period of unprecedented business activity. Yet, in spite of this prosperity, the rich and the poor remained far apart in their attitudes on the role that the government should take in the economy. For example, during the postwar boom, the question of government control and ownership of industry came to the fore. This time the general housing shortage drove many, especially among the poor, to consider the merits of government-sponsored housing, or allocation of the scarce materials which go into construction. The following breakdowns show that faith in the ability of private industry to cope with the shortage dwindled according to the degree of prosperity of the group considered:

April 1946—Do you see the present housing shortage as a problem that industry, if left pretty much alone, would be able to work out itself, or as a problem that won't get straightened out until the government does a lot more than it has?

	Nat'l Total	Prosperous	Upper Middle	Lower Middle	Poor
Leave it to industry	39%	63%	55%	39%	19%
Government help	48	29	35	49	57
Don't know	14	8	10	12	24

In 1948, the public rated the question of prices as the most important issue of the day. In the thirties, the price problem was considered to revolve about deflation. In the late forties, it was a question of inflation. But at both times, regardless of

ECONOMIC DISCONTENT

the contrasting circumstances, the desire for government intervention to remedy the situation increased with deterioration in economic status:

June 1948—Do you think the national government should take steps to do something about lowering prices, or do you think we should let the laws of supply and demand take care of prices without government action?

	Nat'l Total	Pros- perous	Upper Middle	Lower Middle	Poor
Government action	51%	35%	44%	50%	63%
Law of supply and demand	40	63	50	41	23
Don't know	9	2	6	9	14

The boom after the Second World War also saw a period of unprecedented employment. The goal set forth for 60 million jobs was even surpassed at the close of the forties. Yet the public had not forgotten the breadlines of a decade earlier and remained convinced that government must step in to help the destitute:

November 1948—Do you think the government should provide for all people who have no other means of obtaining a living?

	Nat'l Total	Pros- perous	Upper Middle	Lower Middle	Poor
Yes	73%	65%	66%	73%	84%
No	19	26	26	18	10
Don't know	8	9	8	9	6

The way economic classes divided on these issues showed clearly that without exception, the poorer the group, the greater the proportion favoring increased government participation in the running of the economy. The breakdowns also showed that the poorer the group the greater the proportion who were undecided or expressed no opinions at the time these questions were asked. People who answer "Don't know" in surveys are important. They may be withholding their opinion through shyness, inarticulateness, or lack of understanding. Such factors are imponderable, but they cannot be eliminated from consideration. At some later date persons who say "Don't know" may make up their minds on the issues of the day. There is no certain way of telling how they will align themselves. Nevertheless, these answers and breakdowns can be considered in terms of those who *did* express their opinions.

If only the answers of those who expressed opinions were considered, the general trend was the same, except that the gap between the economic classes became slightly greater. For instance, in the 1939 survey on unemployment, 41 per cent of the prosperous felt it was the function of government to step in as against 75 per cent of the poor who felt likewise. On wealth redistribution, 18 per cent of the prosperous as against 53 per cent of the poor voted favorably. Five per cent of the rich and 20 per cent of the poor wanted all the insurance companies to be government-owned.

In 1948, 63 per cent of the lowest economic group, as against 32 per cent of the prosperous group, welcomed government aid for the housing industry. Contrary to the trend shown in answers to other questions, the poor included the smallest proportion of people undecided on the proposition that the government should provide a living for the unemployed.

It has been the dream of farmer-labor parties through the years to draw together the discontented from both the cities and the farms. Yet those who work on the farms and those who work in the factories have widely differing interests and ideologies. Is there any common denominator of these two great occupational groups on which a minority party can base an appeal? The breakdowns in Table 3 throw additional light on the general question as to which groups were the most interested in changing the economic *status quo* during the depression of the thirties.

In Table 3, the salaried executive group includes top

ECONOMIC DISCONTENT

officials of industry, managers, and heads of departments, in fact, those who lay out and direct work of others and who make important policy decisions in the running of their business, whether large or small. The farmer is here classified as anyone who owns and operates or rents and operates a farm,

TABLE 3

OCCUPATION AND OPINION ON THE ROLE OF GOVERNMENT IN ECONOMIC AFFAIRS *
(June, 1939)

	Sal. Exec.	Farmer	Farm Hand	Factory Worker	Unemployed	Nat'l Total
THE GOVERNMENT SHOULD:						
Guarantee employment						
Yes	37%	55%	71%	74%	75%	61%
No	57	37	23	18	22	32
Don't know	6	8	6	8	3	7
Regulate utilities						
Yes	46	48	47	53	53	49
No	49	43	32	34	34	39
Don't know	5	9	21	13	13	12
Redistribute wealth						
Yes	20	35	37	49	39	35
No	75	56	44	39	47	54
Don't know	5	9	19	12	14	11
THE GOVERNMENT SHOULD OWN AND OPERATE:						
The factories						
All	2	7	16	10	18	7
Some	5	12	16	16	16	14
None	90	74	49	67	58	71
Don't know	3	7	19	7	8	8
Insurance companies						
All	6	15	22	16	25	13
Some	13	14	15	18	17	14
None	77	62	41	53	52	62
Don't know	4	9	22	13	8	11
The railroads						
All	14	21	34	25	33	22
Some	7	14	17	15	16	12
None	74	58	27	49	42	52
Don't know	5	7	22	11	9	14

* See pages 139 and 140 for full text of questions.

except share-croppers. Thus, this classification comprises a wide variety of people, both rich and poor. The factory worker group includes all those workers below the rank of foreman. The farm wage earners comprise hired help on the farm. In this group are included share-croppers. Finally, the unemployed are those who were out of work at the time they were interviewed, regardless of what their former occupation or status may have been.

With a single exception, the answers show that a greater proportion among the farm hands, factory workers, and the unemployed favored increased governmental responsibility than was the case among farm owners and executives. The farm-hand and the unemployed groups were particularly close together and frequently took the most extreme view in answering these questions. The farm-owner group appeared more moderate, dividing on these issues in much the same proportion as did the general public. This alliance in ideas between the wage earners of the fields and the factories is one which persists through the years in spite of changes in the economy. A question on unemployment, asked in 1947 near the peak of the postwar boom, shows again the same line-up:

October 1947—If it looked like we were going into another depression that would bring large-scale unemployment, what do you think the government should do: a) See to it that people don't go hungry, but let business and industry take the lead in solving the problem of unemployment. b) Take full responsibility for seeing to it that there are enough jobs to go around and take whatever steps are necessary to accomplish this?

	Nat'l Total	Prof & Exec.	Proprietor	Farm Owner	Factory Worker	Farm Hand
Industry take lead	43%	65%	57%	55%	34%	30%
Government responsibility	45	29	35	39	58	55
Don't know	12	6	8	6	8	15

ECONOMIC DISCONTENT

Once the really needy were taken care of, the general public split about evenly as to whether industry or government should cope with the problem.

During the war the government took over or regulated business on a scale never before envisioned in this country, even during the depression years. Yet, in spite of the successful conclusion of the war, a substantial plurality of the general public felt that government regulation had been carried too far:

November 1948—There used to be little regulation of business by the government, but as the years have passed the government has regulated business in a number of ways. Generally speaking, do you think the government has now carried regulation of business so far that it is hurting our economy, or do you think we have about the right amount of regulation of business, or do you think we need even more government regulation of business than we have in order to protect the public?

	Nat'l Total	Prof & Exec.	Pro- prietor	Farm Owner	Factory Worker	Farm Hand
Carried too far	37%	51%	50%	49%	32%	31%
Right amount	27	24	22	28	26	35
Need more	21	19	20	14	32	17
Don't know	15	6	8	9	10	17

Among the occupations, some 50 per cent of the executives, professionals, the proprietors (owners of small, unincorporated businesses) and the farm owners felt that government regulation had hurt business, while the farm and factory worker groups felt far less strongly on that subject. But while the factory and farm workers together were the least worried that government regulation had been carried too far, the farm hands were relatively cool to the idea that more controls were needed.

The polarization of the farm hands and factory workers at one end with the farmers, executives, and small businessmen on the other end of attitude scales measuring opinion on the role of government is of great significance to farmer-labor parties. In the days of the Populists, it was the farmers who spearheaded the agitation for reform. In the twentieth century, even during a depression, the farmers appeared more moderate, taking sides in some cases in much the same manner as the general public and at other times siding closely with the business and executive group.

One explanation of this trend is that farming has increasingly become more of a business than a way of life. The commercial farmer has gradually cornered more and more of the total agricultural market. It has been estimated, for instance, that consolidation of farms and acreage has proceeded so extensively that 50 per cent of the farms now produce 90 per cent of farm products sold in the open market.[1]

The commercial farmer is a proprietor. He hires farm workers and is accustomed to think of labor as a factor in the total cost of production. In short, big or small, a farmer who is a proprietor pays property taxes, hires labor, and wants high farm prices. In contrast, the objective of labor is high wages, low food prices, and a preference for property rather than consumers' taxes. The conflict of interests is further sharpened by the feeling of many farmers that high wages mean high prices for necessary articles needed on the farm.

Yet the emergence of the commercial farmers has but accentuated the plight of the smaller farmers who produce mainly for their own consumption. The main aspect of the lives of these non-commercial or subsistence farmers, particularly during depressions, is poverty. In the heart of the depression some 40 per cent of the farm families in this country had incomes so low that they often could not afford minimum

[1] M. R. Benedict, "Agriculture as a Commercial Industry Comparable to Other Branches of the Economy," *Journal of Farm Economics*, XXIV (1942), 476–496.

ECONOMIC DISCONTENT

material necessities for the maintenance of vigorous physical health.[2] These subsistence farmers could represent, in terms of numbers and thus in terms of votes, an important factor, at least potentially, in politics.

The fact that farm hands, factory workers, and the unemployed together form the groups most interested in the various economic reforms discussed here, is not particularly surprising. The life of a farm hand is an economically insecure one. While some in this group are regularly employed and can expect gradually to work up to owning their own farm, many others are seasonal workers getting work when they can, or following the crops from region to region as the weather changes. The insecurity of the farm hand has been intensified by the mechanization of agricultural production. Thus, for instance, in the South, the increased use of the machine in the fields has meant that farm labor has been chiefly employed only in harvesting, a development which has made for violent seasonal fluctuations in the demand for labor.

There has also been considerable interchange between those working on the farms and those working in the factories. Particularly during times of depression, many of the unemployed in the cities have sought work on the farms. In reverse, there has been a long-term migration from the overcrowded farms of those seeking employment in the factories of the cities and towns. Thus, there is much common background among these three groups.

Among the farm hands the proportion of "Don't knows" was very high, in many cases amounting to a fifth of those interviewed. When the "Don't knows" were statistically eliminated from these breakdowns of the 1939 survey, the farm hands and the unemployed turned out to have nearly identical re-

[2] O. V. Wells, "Agriculture, an Appraisal of the Agricultural Problem," in "Farmers in a Changing World," *Yearbook of Agriculture, 1940* (Washington: U.S. Dept. of Agriculture), p. 388.

sponses. The two groups did not vary from one another in their approval of any of the various measures by more than one per cent. The general trend remained the same, both in 1939 and 1948. A sizable gap existed between the opinions of the executives and the farmers, on the one hand, and the factory workers, unemployed, and farm hands, on the other hand.

Another survey published in February and April, 1943, sheds light on the question as to which groups within labor and the farm population were the most economically discontented. A specially constructed labor and farm sample was used. The labor sample consisted of four subsections of the working force: factory labor (all ranks below foremen), miners, transportation and utility workers, and personal service workers (janitors, beauticians, domestics, laundry workers, *et al*). The farm sample consisted of male farmers responsible for the land they cultivated, including owners as well as tenants and share-croppers. An additional sample was constructed of farm hands.

When answers to questions testing attitudes toward government ownership were statistically broken down, some interesting facts emerged. Among labor, mine workers and Southern workers were most favorably inclined toward government ownership of the industries concerned. Among those in the farm group, tenants, share-croppers, and farm hands were strongest in advocating government ownership. The poorer the farmer, the more he favored nationalization of such enterprises as automobile manufacturing, electric light companies, and packing-house companies.

On the subject of wealth limitation, farmers in general were less egalitarian than wage earners. Breakdowns show little differences of opinion among occupations, especially in the labor group. On the other hand, if only those with opinions are considered, the miners emerged as the most extreme group. Of the miners with opinions, 62 per cent favored in-

ECONOMIC DISCONTENT

come limitation, while some 52 per cent of the other workers' groups favored it.

February and April 1943—When the war is over, do you think it would be a good idea or a bad idea for us to have a top limit on the amount of money any one person can get in a year?

LABOR SAMPLE

	Mine Workers	Factory Workers	Transportation & Public Utility	Personal Service
Good	51%	47%	49%	46%
Bad	31	42	43	41
Don't know	18	11	8	13

FARM SAMPLE

	Farm Owner	Tenant	Share-cropper	Prosperous Farmers	Fairly Well-Off Farmers	Poor Farmers	All Farmers
Good	36%	39%	37%	31%	39%	49%	37%
Bad	54	47	40	63	53	35	51
Don't know	9	14	23	5	8	16	12

If those without opinions are eliminated from the farm group, only 39 per cent of the farm owners as against 45 of the tenants and 48 per cent of the share-croppers favored the proposition. However, when the financial status of the farm group was considered, it showed clearly that the poorer the farmer, the more extreme his views. The poor farmer, in fact, had, roughly, as great a proportion favoring wealth limitation as did the labor group. Antagonism between these two occupation groups, however, was brought out in a question in the same survey in which farmers were asked whether they felt the wages of factory workers were too little, about right, or too much. Only 2 per cent of the farmers felt that workers were getting too little and almost 50 per cent said they were getting too much. Even hired hands, themselves recipients of wages, had little sympathy with labor's fight for higher pay. Economic breakdowns showed that 61 per cent of the prosperous farmers felt wages were too high, while only 35 per cent of the poor ones shared this view.

A similar question, asking the labor sample how it felt about the amount of money farmers were making, showed that there was remarkably little hostility among the labor groups toward the newly found prosperity of the farm group. Some 40 per cent of the factory workers felt that farmers were making too little money, while only 7 per cent felt that they were making too much. This sentiment was shared equally strongly by the other groups included in the labor sample.

In addition, farmers tended in general to take a dim view of the increased strength of organized labor. Only 2 per cent of the farm sample felt that labor unions had done an excellent job for this country and that they should be given more power than they had. More than 50 per cent felt that although labor unions had done some good in the past, they had gone too far and should be watched closely. Another 13 per cent declared flatly that unions were a bad thing and should be done away with. The different groups within the farm sample showed little variation in attitudes toward unions.

Even so, when the chips were down, farmers found it possible to conceive of farmer-labor cooperation:

April 1943—Suppose in five or six years it became clear that Congress was going to be dominated either by labor or by big business interests, and farmers couldn't do anything about it except throw their support one way or the other, which would you want farmers to support?

	All Farmers	Farm Owners	Share-croppers	Tenants	Economic Level HIGH	LOW	Hired Hands
Labor	45%	41%	48%	52%	35%	56%	59%
Big business	25	30	13	17	39	13	13
Don't know	30	29	39	31	26	31	28

While farmers were skeptical of organized labor, in this hypothetical situation they were even more hostile to big business. A majority of tenants, farm hands, and poor farmers clearly preferred labor to big business.

The pollers then asked the labor sample whether it would

like to see farmers as a group more powerful in this country or less powerful. At least 50 per cent of each group, the miners, the factory workers, those in utilities and transportation and personal service, said that the power of the farmer should be increased. It might be noted however, that labor in the South and the Northeast had the greatest proportion favoring more power for the farmer, while the wage earners living nearer the great Western farm belt were cooler to the idea.

To summarize: There were substantial pluralities and occasional majorities which favored many of the measures started under the New Deal aegis. Both in 1939 and in 1948, feeling was particularly high on the subject of unemployment. By implication, this probably means that in the future no party or government can successfully adopt a do-nothing policy in the face of any possible large-scale unemployment. In addition, the public turned a cold shoulder to the more drastic proposals for nationalization of private industry.

There was a considerable diversity of opinion expressed on these matters among the various groups in the population. Without exception the lines of demarcation followed the lines of economic status. The poorer the group concerned, the greater the proportion of that group which favored nationalization and other measures intended to extend government's role in the economy. It was also the case that the poorest groups had the greatest number of persons without opinions. In particular, the farm hands had a high ratio of "Don't knows."

The farm owners and operators, once the torch bearers of economic radicalism in the days of the Grangers and Populists, now appeared more moderate. Proportionately, they were as little interested in the extension of government control over the economy as were the business groups.

Farm workers, wage earners, and the unemployed constituted the groups most sympathetic to increased government control. Within the labor group itself, the mine workers generally showed the highest percentages in favor of extending the realm of government activity. Share-croppers and tenants expressed themselves as being more in favor of extending such government activity than did the farm owners. But among the farmers the poorest groups, regardless of whether they were tenants or owners, showed the highest percentages in favor of increased governmental control.

The unemployed, economically displaced persons who have the least at stake in our system, frequently emerged as the most radical group of all. Together with farm and industrial wage earners and poor farmers, the unemployed form the greatest voting potential for parties of protest—parties of the farmer-labor type.

9

THE COMPONENTS OF PROTEST VOTING: POLITICAL DISCONTENT

REFORM AS ADVOCATED by farmer and labor parties has been essentially two-pronged. On the one hand, these parties have advocated changes in our economic system; on the other hand, they have prodded the major parties into taking stands on important political issues.

We have analyzed the sources of economic radicalism—for the purpose of this study defined in terms of the answers to the preceding questions. What are the sources of political radicalism? More specifically, are those persons who are economically discontent synonymous with those who are politically discontent? What is the extent of dissatisfaction with the two major parties? Are those who are dissatisfied ready to support a farmer-labor party? This chapter is devoted to a consideration of these problems.

In 1938, Philip LaFollette, son of the 1924 Presidential candidate, attempted to revive the progressive movement on a national scale. His party could have been the first real challenge to the two-party system since 1924, when his father rolled up enough votes to hold the balance of power in many states. But the new movement in 1938 failed when independent parties in states outside Wisconsin failed to give it their support.

The following question shows public reaction in 1938 to the proposal for a realignment in the party system of this country:

August 1938—What parties would you like to see competing in the next presidential race: Republicans and Democrats only; Republicans, Democrats and minor parties as before; Republicans, Democrats, and a new, strong third party, or two new parties with all conservatives voting together and all liberals voting together?

	Nat'l Total	Pros- perous	Upper Middle	Lower Middle	Poor
Republicans and Democrats only	44%	52%	48%	42%	42%
Parties as before	21	18	20	22	22
Strong third party	13	11	13	15	11
New coalition	6	6	8	6	5
Other	1	1	—	—	—
Don't know	15	12	11	15	20

The largest group preferred its politics without any competing minority party. Only a small percentage of those interviewed approved of a realignment of the major parties so that all the liberals were on one side and the conservatives on the other. Interestingly enough, there was little difference of opinion among economic classes on the desirability of a strong third party, or a new realignment of the present parties. But the poorer the group, the less it favored the pure and simple two-party system.

Shortly before the 1944 Presidential election, when President Roosevelt was breaking all traditions in running for a fourth term, the public was asked:

October 1944—Do you feel that the Republican and Democratic parties mainly stand for the same things, or that they stand for quite different things?

	Nat'l Total	Prosperous	Upper Middle	Lower Middle	Poor
Mainly the same	46%	54%	52%	48%	39%
Quite different	42	41	41	43	44
Don't know	12	5	7	9	17

Here, almost as many people in the general public felt that the major parties were the same as that they were different. The poorer the group concerned, the greater the percentage

POLITICAL DISCONTENT

of people without opinions. Perhaps for this reason, the prosperous group included the greatest proportion of those who felt that the major parties were the same.

The answers to this question might have led to a belief that dissatisfaction with the major parties was quite widespread. In order to test this sentiment, the following question was asked:

October 1944—On the whole, how do you feel about the present set-up of the political parties here in the United States: Do you find that you are usually satisfied with the stands taken by one or the other of the present big parties, or would you like to see a strong new party entirely different from either of the present parties?

	Nat'l Total	Prof. & Executive	Minor Salaried	Farm Owner	Farm Hand	Factory Worker
Usually satisfied	78%	78%	74%	80%	79%	83%
New party	14	19	22	15	3	12
Don't know	8	3	4	4	18	5

The above answers are interesting because they show little indecision on the subject of the American party system—among the occupations shown here, only the farm hand group had more than five per cent who "didn't know." These answers also show strong satisfaction with the party system as it existed, with the factory workers contributing the highest proportion in agreement with the stands taken by the two big parties. Differences between the occupations were not large enough to be significant, however. And again, in contrast to other findings on economic issues, the poor were as satisfied as the rich with the existing political system:

	Nat'l Total	Prosperous	Upper Middle	Lower Middle	Poor
Usually satisfied	78%	80%	78%	78%	78%
New party	14	18	17	15	12
Don't know	8	2	5	7	10

These two questions, one of which was asked in the depth of the depression when the Roosevelt reform administration was in power, the other of which was asked in the middle of World War II, showed an overwhelming satisfaction with the political system as it existed during that momentous decade of our history.

The fact that the poor people of this country and the wage earners have not interested themselves to any great extent in the establishment of a party catering specifically to their own needs goes a long way toward explaining why class politics has failed at the polls. When a proposal for a new labor party was put squarely up to the house of labor itself, only small percentages favored the idea. The following question was asked of a specially selected sample representing a cross section of the gainful working force of this country, exclusive of proprietors, officials, executives, and professionals:

June 1940—Do you think labor unions should form a national party in the United States in addition to the present parties; or should labor unions support one or the other of the major parties; or should unions as such keep out of politics altogether?

	Labor Total	Miners	Factory	Rail-roads	Other Transp.	Farm Hands	Unemployed
Labor party	11%	16%	15%	10%	14%	10%	11%
Major parties	18	20	18	29	18	10	17
No politics	57	54	55	55	61	57	51
Don't know	14	10	12	6	7	23	21

Clearly, labor in general did not assume an attitude of independence toward the major parties. A majority of labor as a whole wanted labor to keep out of politics entirely. Only a small proportion wanted independent action. The occupational breakdowns show that miners, factory labor, and some transportation workers, the groups most desirous of changing the economic *status quo,* were those who had the greatest proportion desiring a labor party. The farm hand and the unemployed groups, however, usually the highest in expressing economic "radicalism," were here relatively cool to the idea of a labor party.

POLITICAL DISCONTENT

But within labor, there were faint stirrings of political rebellion. When the attitudes of CIO members were isolated, it was evident that there were as many supporters of a new labor party as of the traditional parties. Union membership also appeared to be a factor in determining the extent of sympathy with independent political action. Both AFL and CIO members contributed a higher proportion of people wanting a labor party than did non-union members:

	Labor Total	CIO	AFL	Non-Union
National labor party	11%	24%	19%	8%
Support major parties	18	23	23	16
Keep out of politics	57	46	51	60
Don't know	14	7	7	16

In 1943, the CIO's Political Action Committee was formed. It represented labor's most significant attempt in a generation at political action on a national scale. Claiming to be nonpartisan in character, it operated within the two-party system and threw its support aggressively to candidates it favored regardless of their political affiliation.

The possibility that it could control the votes of the tremendous membership of the CIO made it a real challenge to the two major parties. From its very start the PAC was a controversial organization, being attacked both from within and from without the labor movement. Yet it represented little that was new in political methods. The Farmers' Non-Partisan League, which reached the height of its power in the 1920's, and Labor's Non-Partisan League, which operated during the 1930's, proceeded on a generally similar basis. Three years after PAC was formed, *Fortune Survey* sampled public opinion on the role unions should perform in politics.

Only a small proportion of the population favored the idea that unions should organize their own party. Significantly, the size of the group having no opinions on this important question again increased sharply as economic status deteriorated:

November 1946—Which do you think American labor unions should do: Support candidates put up by one of the present political parties, form their own labor party and run their own candidates, or keep out of politics altogether?

	Nat'l Total	Prosperous	Upper Middle	Lower Middle	Poor
Labor should form own party	12%	9%	10%	13%	11%
Support present parties	21	25	26	21	16
Keep out of politics	49	58	54	51	41
Don't know	18	8	10	15	32

	Prof.	Sal. Exec.	Farm Prop.	Other Prop.	Factory Worker	Farm Hand
Labor should form own party	13%	8%	13%	12%	16%	6%
Support present parties	37	31	13	21	22	17
Keep out of politics	41	58	54	61	44	46
Don't know	9	3	20	6	18	18

In the poorest group, there were more people without opinions than there were people who felt that organized labor should enter politics in some form, either by starting its own party, or by supporting the present parties. If only those people having opinions were considered, interest in a labor party increased among the poorer groups: 10 per cent of the prosperous and 11 per cent of the upper middle class felt that a labor party would be a good idea, while some 15 per cent and 16 per cent of the lower middle and poor classes thought labor should form a party. But in reality, it is not possible to ignore this large group of "Don't knows." The poor usually contribute a relatively large proportion of people without opinions, regardless of the subject matter or the timing of the question. These people in the poor group in many parts of the country are presumably precisely the persons whom the great machines of the major parties have been so successful in organizing.

More than half of the executives, farm owners, and proprietors felt that labor should stay out of politics entirely. Sub-

POLITICAL DISCONTENT 161

stantial pluralities among labor itself agreed to this proposition.

On the specific question of the PAC, the country divided as follows:

November 1946—The Political Action Committee of the CIO (usually called the PAC) has been active all over the country in supporting candidates for political office that labor approves of and opposing those labor disapproves of. On the whole, is the PAC the kind of organization you would like to see continued or not?

	Nat'l Total	Professional	Sal. Exec.	Farmer	Proprietor	Factory Worker	Farm Hand
PAC should be continued	18%	23%	16%	12%	15%	31%	15%
Not continued	49	67	72	60	74	37	37
Don't know	33	10	12	28	11	32	48

The factory workers and the professionals were the only groups in which more than a fifth favored continuation of the PAC. The latter category consists mostly of doctors, lawyers, artists, musicians, and teachers. Before Congress in 1943 placed certain restrictions upon the spending by unions in political campaigns, the professional group supplied a good deal of the leadership of the PAC.

The farm operator group also showed itself to be politically akin to both the executive class and the proprietors. This identification became even closer when only those people who expressed opinions were considered. Like his employer, the farm hand was cool to the PAC.

Understandably, both farmers and farm hands had large numbers of people who preferred not to express an opinion on the PAC, an organization which normally confined its activities to labor in the cities. But the high degree of inarticulateness as expressed by the absence of opinion among the factory workers is surprising.

Here also, as in the earlier survey shown in these pages,

membership in a union appeared closely tied in with increased approval of labor's participation in politics:

	Nat'l Total	Union Member	UNION STATUS: Union Member in Family	No Union Member in Family	MEMBER OF: CIO	AFL
PAC should be continued	18%	43%	23%	14%	58%	33%
Not continued	49	36	38	53	24	46
Don't know	33	21	39	33	18	21

The gap in sympathy toward the PAC between those affiliated with a union, those not directly connected with one, and those who by implication were removed from unions, was considerable. It remained large even if those people without opinions were statistically eliminated from the breakdowns. Significant also was the relative interest displayed by CIO members in their own organization.

In 1948, a new attempt on a national scale was made to organize an independent party which would appeal to farmer and labor interests. Called the Progressive party and headed by two candidates from the farm-belt states of Iowa and Idaho, the new party had many of the outward characteristics of previous farmer-labor parties. But it was clear from the start that it was not to inherit the mantle of the LaFollette Progressives or the Populists. In 1948 the CIO-PAC itself came out against a third party. It was therefore not surprising that the Wallace Progressives set no records at the polls in 1948. Their vote of 2.4 per cent of the total vote was not substantially higher than that received by other third parties during the New Deal days. The Socialists in 1932, for example, got 3 per cent of the total vote. The Union party of 1936 received almost 2 per cent of the vote.

Nevertheless, as the first national farmer-labor party to be organized since the death of Roosevelt, the character, organization, and following of the Progressive party provided addi-

POLITICAL DISCONTENT   163

tional information on the nature of farmer-labor parties. Surveys taken in the early days of the Wallace campaign (a campaign which started months ahead of the major party campaigns) showed that the youth of America, persons living in large cities, and members of labor unions were the groups most favorably disposed toward the leader of the Progressive party.

The source of Wallace support became more apparent when those persons who said that they preferred the Progressive candidate were examined at closer range. A *Fortune Survey* published in June, 1948, revealed the following results: 53 per cent of those who preferred the ex-farmer from Iowa for president were aged 21 to 34. Only 36 per cent of all voters were in this age group. Of the Wallace supporters, 61 per cent were found to come from cities in the 100,000 or larger class. Nationally, 32 per cent of the total population live in such cities.

Among the various occupations interviewed, factory workers and other non-farm wage earners were the only groups to give Wallace significant support. In proportion to their actual numbers in the population, the housewives of America showed themselves to be the group least enthusiastic about Wallace and his Progressive party. Yet, analysis of the Wallace supporters in terms of occupational background showed that he had attracted some support from every group in the population. A few members of the executive and prosperous classes expressed a preference for the Progressive candidate as did some small businessmen. As the campaign developed, however, much of this early support evaporated.

On the basis of the data so far examined, the following generalizations may now be stated: Those groups showing the highest proportion of "radicalism" in their views on economic issues were frequently not those the most politically "radical," that is, those who desired to challenge the two-party system as it stood during the thirties and forties. There was often quite a gap between the opinions of the poor and of the rich and among the various occupational groups examined. Yet, the great and overwhelming bulk of the citizenry, despite their economic beliefs, subscribed to the present two-party system.

Within the small percentages of the population which expressed a desire for a new party or for a labor party, certain patterns were evident. Generally speaking, the more prosperous and established groups took the more sophisticated point of view and saw little difference between the two major parties. The dissidents among these groups favored a third party presumably of a middle class type. During the decade observed in these pages, such a party probably was intended as a challenge to the Democratic administration.

In answer to the more specific proposal for a labor party, class lines were still not so clear as they were on economic subjects. Generally, it can be said that the poorer the economic status of the group, the higher the proportion favoring such a party. This relationship between economic and political reformism became clearer when only those people having opinions were considered. Attitudes on political matters presented a contrast to answers on economic matters, where lower economic status invariably meant a substantially higher ratio favoring economic radicalism.

Labor, especially organized labor, had a consistently higher proportion favoring a party for itself. But farm hands, usually the strongest in advocating economic change, were cool as a group toward the idea of a labor party or of labor participation in politics. Farmers, once the backbone of earlier radical movements, had about the same proportion of persons sympathetic to the idea of a labor party, or to the demands of such a party, as did small businessmen.

In view of the foregoing generalizations, what are some of the factors which might explain the inability of farmer-labor parties to attract a more sizable following, particularly among those very groups which an economic determinist would expect to be drawn to their program?

A factor of significance is that most Americans consider themselves to be members of the middle class and therefore are wary of parties which clarify class lines. In fact, in 1940, 79 per cent of the public, when questioned by *Fortune*, classified themselves as belonging in the middle class, while only 16 per cent—split about evenly—put themselves in the upper or lower classes. This strong identification with the middle class was consistently held by every geographical or economic group. Thus, 75 per cent of the prosperous groups and 70 per cent of the poor group classified themselves as members of the middle class.[1]

Inertia and voting habits also probably work against the success of farmer-labor parties. In October, 1944, *Fortune* asked with which party the respondents usually identified themselves. The *Survey* then compared these answers with answers to the questions: "So far as you know, would you say your father was mainly a Republican, Democrat, or what?" The following similarities in political allegiance between male parents and their offspring then developed:

70 per cent of the people whose fathers were Republicans were themselves Republicans;
75 per cent of those whose fathers were Democrats were themselves Democrats;
53 per cent of those whose fathers were independents were themselves independents.

A third important factor which emerges from this examination of public opinion polls is that of apathy. Included in this category must be all those who express no opinions at all. Among the various occupations, farm hands, factory workers, and the unemployed usually have the highest proportion of

[1] See *Fortune* magazine, XXI, No. 2 (February, 1940), 14, 20, for full text and answers to questions.

"Don't knows." In the farm-hand group particularly, at least a fifth were without opinions.

Finally, it can also be said, without exception, that lack of opinion increases in proportion to the decline in economic status. Thus it can be seen that the groups to which farmer-labor parties are trying to appeal are the very groups which have a high ratio of inarticulateness, at least when dealing with political questions. Answers to a question published in February, 1948, illustrated this clearly. Without describing them, the *Fortune Survey* asked the public whether it had heard of a number of organizations taking an active interest in politics. Among the organizations were the Political Action Committee of the CIO and the National Association of Manufacturers. Some 50 per cent of the people interviewed had never heard of either organization.

These answers showed a tremendous lack of political awareness on the part of a great number of American citizens and gave an idea of the nature of one obstacle which farmer-labor parties must overcome. When the answers to the question on the Political Action Committee of the CIO were broken down by the educational level of respondents, they gave these results: Some 78 per cent of those who had a college education had heard of the PAC, while 50 per cent of those who had been to high school had also heard of it. In contrast, only 30 per cent of those who had eighth grade or less education had heard of this controversial organization. This is but one bit of evidence of the influence of education in opinion formulation and it shows clearly a reason for the lack of political awareness among the less privileged groups.[2]

Inability to form an opinion is important, because this inability is a factor in determining whether a person takes the trouble to vote. A study of the 1940 elections based on

[2] For additional material on opinion formulation and education, see *Fortune*, XXXII, No. 4 (October, 1945), 282, 285, 286, 288.

surveys of the National Opinion Research Council came to this conclusion on opinions and voting: "The percentage of non-voters without opinions is roughly double that of the voters." [3] The same study added that "the more economic security and education a person has the more likely he is to participate in elections." [4] It was further found that while 81 per cent of those with a college education voted, only 61 per cent of those with a grade school education did so. Some 68 per cent of those in the middle economic group voted, as against 53 per cent of those in the poor economic category.[5]

From the foregoing analysis it is clear that farmer and labor parties have faced formidable psychological barriers in their efforts to attract voters. While such barriers are probably not inflexible, they have thus far in our history helped to prevent the formation on a permanent basis of a strong national farmer-labor party.

[3] Gordon M. Connelly and Harry H. Field, "The Non-Voter, Who He Is, What He Thinks," *Public Opinion Quarterly*, VIII (1944), 182.
[4] *Ibid.*, p. 179.   [5] *Ibid.*

10
PERSPECTIVE

FARMER-LABOR PARTIES have fulfilled two leading functions—to popularize ideas and to act as vehicles for discontent. As popularizers of ideas and issues these parties have been most successful. They have called attention to important problems which the major parties have seemingly ignored. They have excited public interest in reform measures. Ironically, at the moment of greatest success—adoption of a principal issue by a major party—the farmer or labor party tends to die. Ideological and electoral success rarely coincide. This is a recurrent pattern of American politics.

Farmer-labor parties also serve as vehicles for discontent. By their existence they make it possible for discontent to be registered at the polls when the major parties ignore key issues. Protest voting of this type is therefore a permanent feature of the American political landscape. So long as we have free elections and preserve our traditional civil liberties, farmer-labor parties may be expected to challenge the major parties, and thereby to strengthen the democratic process.

Protest voting may ebb and flow, but it never completely dies out. In contrast to the major parties, individual farmer-labor parties are relatively short-lived. Nevertheless, the persistence of farmer-labor parties has been impressive. Since 1872 at least one party of this type has been represented in every Presidential election. Yet the individual party itself is of no great importance. What apparently happens is that under certain conditions many voters turn to the most acceptable minor party of the day as a vehicle for registering their own discontent. Thus, the Granger, Greenback, and Populist parties attracted similar support. The Socialist party furnishes another case in point. After the passing away of the Populists, the Socialist party received the largest share of votes cast by the discontented for protest parties in the South and the Southwest.

It is of great practical significance that the extremist parties have never done very well. Conversely, those groups with broad popular followings have on occasion made excellent showings. In this group must be included the Grangers, Greenbackers, Populists, Minnesota Farmer-Laborites, Wisconsin Progressives, and the American Labor and Liberal parties of New York State.

It is not possible to find any exact relationship between protest voting and the condition of the economy. Yet, historically, farmer-labor parties have attracted their greatest vote when the economy was going downhill but before rock bottom was reached. Some time before the bottom was reached, a major party always absorbed most of the potential protest vote which might otherwise have gone to a farmer-labor party. The flexibility of the major parties—which has been frequently illustrated in this study—is an outstanding characteristic of our political system.

There have been considerable long-term and basic shifts in the locale of protest voting of the farmer-labor type. The absolute vote for farmer-labor parties is shifting, under the sheer pressure of population, to the most heavily populated regions. Since 1900, when farmer-labor parties have analyzed their votes and membership, they have found that a good proportion of both came from the nine states which form the heart of industrial America.

But on a per capita basis the picture is somewhat different. Before the advent of the New Deal, and particularly in the great show of farmer-labor voting strength in 1924, the greatest strength of farmer-labor parties, man for man, was in the Great Lakes and Far West regions. It should be noted, however, that the slow but consistent trend toward increased voter interest in parties of this type in the urban East goes back to the start of the century.

During the days of Franklin Roosevelt's administration there was little farmer-labor voting except in Wisconsin, Minnesota, and New York. After the end of the war, the first stirrings of political protest on a national scale began. The centers of these movements were in the cities of the industrial East and in the Great Lakes region as well as in the Far West. The Progressive party of 1948 was to some extent symptomatic of the contemporary political unrest. More significant were intimations of Walter Reuther, head of the United Automobile Workers, and some Railroad Brotherhood and AFL leaders, that labor might organize a national political party. The results of the 1948 election quieted, at least temporarily, reflections along this line.

In contrast to many labor leaders, farm leaders have shown relatively little interest in recent years in protest parties. However, this lack of interest does not mean that farm protest voting is necessarily a thing of the past. Within the farm population itself there is a tremendous sympathy for the type of economic changes that farmer-labor parties would like to effect—changes such as wealth equalization and greater governmental control over the economy. The farm workers in particular (who comprise some 40 per cent of those gainfully employed on farms) show great interest in these demands.

Like other groups, farmers show increased interest in the kind of program put forth by parties of protest as their own economic status declines. Historically, this interest has several times taken the form of considerable support for farmer-labor parties. Especially in the Western agricultural states, where the tradition of political insurgency runs deep, it is likely that an agricultural depression would stimulate voting for parties of the politically dissatisfied. The 1948 Presidential election showed again that even in prosperous times farmers react quickly to what they consider threats to their own economic well-being.

It is not possible to predict whether the political development in this country will lead to a party modeled on the British Labour party.[1] The two-party system in the United States showed itself adequate to meet the challenges posed by the tremendous depression of the 1930's and by the Second World War.

Even if the two-party framework remains, those who are politically discontented with the present major parties have another alternative than the creation of a new farmer-labor party. This alternative is the realignment of the major parties, with the "have-nots" and liberals opposing the "haves" and the conservatives. Should such a realignment occur, much of the *raison d'être* of a strong farmer-labor party would disappear. Such a change, however, is not likely, despite the trend toward polarization of interests within the two-party system during the New Deal era. There remains the prospect of establishing a new party.

The obstacles in the path of such a possible development are enormous. Militating against it are apathy, ignorance, fixed voting habits, legal limitations on third party activity,

---

[1] Comparisons between the British Labour party and a possible farmer-labor party of considerable proportions in this country are frequently apt to be misleading. There is no sizable agricultural bloc as such in British politics. Only 6 per cent of Britons are classified as farmers while 54 per cent are classified as workers. See John C. Ranney and Gwendolen M. Carter, *The Major Foreign Powers* (New York: Harcourt, Brace, 1949), p. 14. This same work contains a description of British parties and elections, pp. 54–90. On the history of the Labour party to 1929, see Ergon Wertheimer, *Portrait of the Labour Party* (New York: Putnam, 1929). For the period to 1937, see Clement R. Attlee, *The Labour Party in Perspective* (London: V. Gollancz, 1937). A recent analysis of the general program of the Labour party can be found in John Parker, *Labour Marches On* (London: Penguin Books, 1947). Current developments are regularly reported in the various publications of the British Labour party.

and legal restrictions which, in effect, disfranchise large groups of the poor in some parts of the country. Yet, given a deteriorating economy and a subsequent lack of faith in the traditional middle-class major parties, the rise of a genuine national farmer-labor party is possible.

The data of this study indicate that agricultural workers, union labor, and the unemployed form the great potential reservoir which might furnish membership to such a party. Professional and intellectual groups might be counted on for considerable financial and moral support, although they could contribute few votes.

Whether a national farmer-labor party eventually develops or not, parties of the farmer-labor type will undoubtedly continue to perform, and to perform effectively, their two great historical functions—popularization of ideas and channeling of discontent. In themselves, these functions are of first-rate importance, not only to the major parties, but also to the American people in the perpetuation of political democracy.

APPENDIX A

Voting for Head of Tickets of Farmer and Labor Parties in Presidential and State-wide Elections, 1872–1948

Year	Farmer-Labor Votes (in Thousands)	Farmer-Labor Votes Expressed as Percentage of Total Vote
1872	29	0.4
1874	195	3.3
1876	82	0.9
1878	870	12.8
1880	309	3.3
1882	333	4.6
1884	175	1.7
1886	307	3.4
1888	148	1.3
1890	452	4.9
1892	1,041	8.7
1894	1,442	12.5
1896	124	1.1
1898	325	3.0
1900	175	1.3
1902	292	2.1
1904	650	4.1
1906	335	3.0
1908	463	3.1
1910	525	4.2
1912	906	6.2
1914	586	4.0
1916	600	3.2
1918	439	3.3
1920	1,200	4.5
1922	884	4.1
1924	4,898	16.8
1926	552	2.8
1928	345	0.9
1930	809	3.4
1932	1,069	2.7
1934	1,529	5.1

APPENDIX A

Year	Farmer-Labor Votes (in Thousands)	Farmer-Labor Votes Expressed as Percentage of Total Vote
1936	1,466	3.2
1938	1,359	3.7
1940	598	1.1
1942	1,276	4.8
1944	949	1.9
1946	741	2.2
1948	1,553 *	3.2

* Preliminary total based on Associated Press reports.

APPENDIX B

PARTIES AND STATES OF FARMER AND LABOR PARTY REPRESENTATIVES IN CONGRESS, 1860–1948

Congress	Number in Senate	Number in House		Number from Each Party and Each State
37th–43d	0	0		
44th	0	3	HOUSE	Independents: Ill., 3
45th	0	0		
46th	0	15	HOUSE	Republican Nationals: Mo., 1; N.C., 1; Pa., 2; Vt., 1; Ala., 1; Ill., 1
				Democratic Nationals: Ill., 1; Tex., 1; Ind., 1; Ia., 2; Me., 2; Pa., 1
47th	1	12	SENATE	Readjuster: Va., 1
			HOUSE	Greenbackers: Me., 2; Mo., 4; Pa., 2; Tex., 1; Ala., 1
				Readjusters: Va., 2
48th	2	7	SENATE	Readjusters: Va., 2
			HOUSE	Greenback Labor: Pa., 1; Ind., 1
				Readjusters: Va., 5
49th	0	2	HOUSE	Greenback Labor: Ia., 1; Pa., 1
50th	0	3	HOUSE	Labor: Pa., 1; Va., 1; Wisc., 1
51st	0	0		
52d	2	9	SENATE	Populist: Kans., 1
				Independent: S.D., 1
			HOUSE	Farmers' Alliance: Kans., 5; Minn., 1; Neb., 2; Ga., 1
53d	3	11	SENATE	Populists: Kans., 1; Neb., 1
				Independent: S.D., 1
			HOUSE	Populists: Col., 2; Kans., 5; Minn., 1; Neb., 2; Nev., 1
54th	4	7	SENATE	Populists: Neb., 1; N.C., 1; Kans., 1
				Independent: S.D., 1
			HOUSE	Populists: Kans., 1; Col., 1; Ala., 1; Neb., 1; N.C., 2
				Silver: Nev., 1

APPENDIX B

Congress	Number in Senate	Number in House		Number from Each Party and Each State
55th	8	28	SENATE	Populists: Neb., 1; N.C., 1; Kans., 1; Ida., 1; Wash., 1
				Silver: Nev., 1; Neb., 1
				Independent: S.D., 1
			HOUSE	Populists: Cal., 2; Col., 1; Kans., 6; N.C., 4; Ala., 1; S.D., 1; Neb., 1
				Silver: Col., 1; Mont., 1; Nev., 1
				Fusion: Neb., 2; Mich., 1; Ill., 2; Wash., 2; O., 2
56th	9	7	SENATE	Populists: Ida., 1; Kans., 1; Neb., 1
				Silver: Nev., 2; S.D., 1
				Progressive: N.C., 1
				Fusion: Wash., 1
				Independent: S.D., 1
			HOUSE	Populist: Col., 1; Kans., 1; Neb., 3
				Silver: Col., 1; Nev., 1
57th	2	5	SENATE	Populist: Kans., 1
				Fusion: Wash., 1
			HOUSE	Populists: Neb., 2; Mont., 1; Ida., 1
				Silver: Col., 1
58th	0	2	HOUSE	Union Labor: Cal., 2
59th–61st	0	0		
62d	0	1	HOUSE	Socialist: Wisc., 1
63d	0	0		
64th	0	1	HOUSE	Socialist: N.Y., 1
65th	0	1	HOUSE	Socialist: N.Y., 1
66th	0	0		
67th	0	1	HOUSE	Socialist: N.Y., 1
68th	2	2	SENATE	Farmer-Labor: Minn., 2
			HOUSE	Farmer-Labor: Minn., 1
				Socialist: Wisc., 1
69th	1	5	SENATE	Farmer-Labor: Minn., 1
			HOUSE	Farmer-Labor: Minn., 3
				Socialist: Wisc., 1; N.Y., 1

APPENDIX B

Congress	Number in Senate	Number in House		Number from Each Party and Each State
70th	1	3	SENATE	Farmer-Labor: Minn., 1
			HOUSE	Farmer-Labor: Minn., 2
				Socialist: Wisc., 1
71st	1	1	SENATE	Farmer-Labor: Minn., 1
			HOUSE	Farmer-Labor: Minn., 1
72d	1	1	SENATE	Farmer-Labor: Minn., 1
			HOUSE	Farmer-Labor: Minn., 1
73d	1	5	SENATE	Farmer-Labor: Minn., 1
			HOUSE	Farmer-Labor: Minn., 5
74th	2	10	SENATE	Farmer-Labor: Minn., 1
				Progressive: Wisc., 1
			HOUSE	Farmer-Labor: Minn., 3
				Progressive: Wisc., 7
75th	3	13	SENATE	Farmer-Labor: Minn., 2
				Progressive: Wisc., 1
			HOUSE	Farmer-Labor: Minn., 5
				Progressive: Wisc., 7; Cal., 1
76th	3	4	SENATE	Farmer-Labor: Minn., 2
				Progressive: Wisc., 1
			HOUSE	Farmer-Labor: Minn., 1
				Progressive: Wisc., 2
				American Labor: N.Y., 1
77th	1	5	SENATE	Progressive: Wisc., 1
			HOUSE	Progressive: Wisc., 3
				Farmer-Labor: Minn., 1
				American Labor: N.Y., 1
78th	1	4	SENATE	Progressive: Wisc., 1
			HOUSE	Progressive: Wisc., 2
				Farmer-Labor: Minn., 1
				American Labor: N.Y., 1
79th	1	2	SENATE	Progressive: Wisc., 1
			HOUSE	Progressive: Wisc., 1
				American Labor: N.Y., 1
80th	0	2	HOUSE	American Labor: N.Y., 2
81st	0	2	HOUSE	American Labor: N.Y., 1
				Liberal: N.Y., 1 *

* Special election, May 17, 1949.

INDEX

Absorption of issues by a major party, 112-14
A.F. of L. *see* American Federation of Labor
Agitational parties, 4
Agrarian parties, *see* Farmer and labor parties
Agrarian socialism in North Dakota, 25
Agricultural Adjustment Adminstration, 21, 85
Agricultural products, prices fluctuate violently, 82
Amalgamated Clothing Workers, 9, 108, 122
American Civil Liberties Union, model election law proposed, 128
American Federation of Labor, 10; attitude toward a labor party on a national scale 159, 170; La Follette supported by, 23, 93, 98, 105; pure and simple unionism, 106 f.; war chest to combat Taft-Hartley Act, 107
American Labor party of New York State, 9, 26, 53, 107 f., 169; balance of power held by, 117 ff.; candidates supported by, 74, 118-21; factionalism, 109; legal restrictions have little effect upon, 125; Liberal Party and, could control N.Y. politics, 122; membership, 121; record at polls, 33 ff. *passim*, 40, 66, *chart*, 63; representation in Congress, 41; rivalry between Liberal party and, 120; strength in N.Y. City, 120
"American plan," 98
Americans for Democratic Action, 40
"American system of finance," 104
Anti-Monopoly party, 7, 17
Apathy of lower economic group, 165
Arnall, Ellis, 131
Arnett, Alex M., 69
Australian ballot, 20

Balance-of-power tactics, 117-23
Ballot, filing fee required, 127; ways of placing party on, 125 ff.
Banking practices denounced, 13
Bank of North Dakota Act, 26
Banks, government ownership, 28
Bargaining position of labor, 1920's, 92; La Follette plank, 24
Bingham, Barry, 131
Blaine, James G., 37
Bleakley, William F., 118
Bonaparte, Charles J., 16
Bourbon Democracy in South, 130, 131
Boyle, James E., 56
Bridgeport, Conn., socialist mayor, 106
British Labour party, 171
Bryan, William Jennings, 7, 20, 111
Buck, Solon J., 86; quoted, 112
Building-trades strike for ten-hour day, 12
Bull Moose party, *see* Progressives of 1912
Burchard, S. D., 37
Business, condition of, and voting for farmer and labor parties, 1872-1948, 89-92, *chart*, 90; *see also* Industry; Monopoly
Business cycles, *see* Depressions
Butler, Benjamin, 37

Capitalism denounced, 22
Carpenters' Union, 108
Chicago labor in politics, 93
Child labor, 17, 21
Chinese immigration, 15, 17
Christensen, P. P., 52, 74
CIO, *see* Congress of Industrial Organizations
Civil Service Commission, 16
Civil service reform, 15, 29
Class legislation, repeal demanded, 18
Class politics, 5; failure at polls, 158; mobilization for action weak, 102;

Class politics (*Continued*)
  shunned, 165; *see also* Negroes; Poor
Cleveland, Grover, 37
Collective bargaining, 24, 92
Colored Alliance, 131
Commodity Credit Corporation, 21
Communication, government ownership of means of, 21; *see also* Government ownership
Communist party, 8, 23, 31; campaign of 1932, 113; laws excluding from ballot, 127; legal status in New York lost, 109, 121; middle class support not solicited, 105
Congressional elections, *see* Elections, congressional
Congress of Industrial Organizations, attitude toward a labor party, 159; Labor's Nonpartisan League dominated by, 9
—— Political Action Committee, 10; aggressive political action, 107, 159; continuation desirable? 161, 162; O'Dwyer's plurality a victory for, 120; Wallace not endorsed by, 117, 162
Conservatism in Northwest, 65
Constitutional Amendments, 17-20 *passim*
Convict labor system, 17, 21
Cooke, Jay, & Company, 96
Cook County, Ill., underrepresentation in state legislature, 42
Cooper, Peter, 102
Cooperative enterprises, 23
Costs, fixed, 83, 85, 88
Cotton, 68
Coughlin, Charles, 9
Courier-Journal, Louisville, 131
Courts, federal: Granger cases, 14; use of injunctions, 20, 22, 24
Cox, James R., 113
Coxey, Jacob S., 113
Credit systems, efforts to correct injustices in, 88
Currency, flexibility in supply, 21
Curtis, George W., 16

Dana, Richard Henry, 16
Davis, Jeff, 70
Debs, Eugene V., 8, 23, 70
Debt, *see* Mortgage debt
Deflation, post-Civil War, 14
De Leon, Daniel, 8
Democratic-ALP-Liberal coalition, 41
Democratic-Farmer-Labor party of Minnesota, 25, 40
Democratic party, absorbed most of protest vote, 1896, 1932, 80, 91; campaign of 1924, 49, 115; Farmer-Labor party candidates in Minnesota supported by, 39; fusion with Populists, 58 ff., 80, 110 ff.; protest vote needed to win Presidential election, 49; Roosevelt Democrats absorbed discontent which followed economic collapse, 68; surrender to demands of minor parties, 113
Democrats, Southern: AFL attack on, 107; battle between Southern Populists and, 130 ff.
Depressions, and effect on voting of migrant workers, 134; groups interested in changing economic *status quo*, 144, *tab.*, 145; history of, since Civil War, 95 ff.; of 1930's, 82, 83, 91, 138; panic of 1873, 14, 15, 96; prices and wages during, 88, 91; stock market collapse, 1929, 98; unemployment, 138
Dewey, Thomas E., gubernatorial elections, 118, 120, 121; presidential, 119 f.
Direct election, 19, 24
Direct primaries, 21
Disarmament, 28
Discontent, economic and political, 30, 137-67; and protest voting, 137-54; farmer-labor parties as vehicles for, 168, 172 (*see also* Farmer and labor parties); peak protest years, 35; shift in centers of, 50 ff.
Disfranchisement, legal restrictions

INDEX

which effect, 172; of persons on relief and paupers, 134; poll tax, 131 ff.
Dissension and factionalism, 108-10
Dubinsky, David, 110

East, protest voting in industrial, 61, 64, 74, 75
Eaton, Dorman B., 16
Economic affairs, government intervention in, 138 ff.
Economic conditions, comparison of emonomic and political factors, 95-100; reflected in voting behavior, 5, 48, 68, 77, 78-101 *with charts*, 169; strikes a symptom of economic discontent, 94
Economic discontent, *see* Discontent
Economy, big business control of, 28; shift from agricultural to industrial, 50 ff.
Education, movement for free, 12, 13, 17, 29
Educational parties, 4
Election ballot, *see* Ballot
Election law, model, proposed, 128
Elections, disfranchising methods, 131, 134 ff., 172; economic conditions reflected protest voting, 5, 48, 68, 77, 78-101 *with charts;* effect of prices upon, 1870-1948, 83, *chart,* 84; record at polls, 32-49; restrictions on voting in South, 129 ff.; union activity and voting, 92-95; voting for head of farmer and labor tickets, 1872-1948, *tab.,* 173 ff.; *see also* Ballot
—— Congressional: candidates of farmer and labor parties elected to House, 1870-1948, 43-47, *chart,* 45; direct, of Senators demanded, 19; economics of protest voting exemplified, 80 ff., *chart* 81; non-ideological parties make better showing than those with rigid philosophy, 45, 46
—— Presidential: direct, proposed, ᴖ4; farmer and labor parties in, 6; foreign-policy questions introduced into, 76; geographic patterns of protest voting, 50-77 *with charts;* third parties in field, 1932, 112; vote for farmer and labor parties, 1872-1948, 32-37, *tab.,* 34, *charts,* 35, 38
—— state: analysis of voting histories, 53 ff.; vote for farmer and labor parties, 1872-1948, 37 ff., *chart,* 38
Electors, Presidential: state requirements, 128
Employment, government guaranteed, 138, 140, 144 ff., *tabs.,* 141, 145
Equal Rights party (Loco-Focos), 6

Factionalism, and dissension, 108-10; and fusion, 123
Factories, inspection of, 17; nationalization, 140, *tabs.,* 141, 145; *see also* Industry
"Farm bloc" of the 1920's and 1930's, 60
Farm Credit Act, 85
Farmer and labor parties, absorption of issues by a major party, 112; balance-of-power tactics, 117-23; continuity of ideas, 27, 30; economics of protest voting, 5, 48, 68, 77, 78-101 *with charts;* election on a national scale never won by, 117; extremist, unlikely to prosper at polls, 49; farm prices and number of candidates elected to Congress compared, 1860-1948, 80-82, *chart,* 81; factors in development of, 3 ff., 30, 76, 101; functions, 168, 172; lack of permanence of individual parties, 47; legal restrictions, 125 ff.; local, established in West, 7; national, 167, 170, 171, 172; organizational difficulties, 114; political polarization, 147 ff.; population trend from rural to urban areas affects, 50 ff., 75; Presidential campaign, 1932, 113; press coverage, 115; reforms traceable to, 12, 29, 155; regional

Farmer and labor parties (*Cont.*)
strength, 46, 169, 170; representatives in Congress, 1860–1948, *list*, 175-77; states controlled by, 117; strategy and tactics, 102-24; strength shifting to industrial areas, 47; support sought from middle class and from local and state reform groups, 10, 105; trends since 1900, 73-75; unity, 103, 108; victories on state level, 39; why unable to attract more sizable following, 164 ff.; *see also* Farmers' parties; Labor parties; *and under names of parties, e.g.*, Progressive party
—— voting: area of consistently high per capita, for, 54-57, *chart*, 55; area of greatest absolute vote for Presidential candidates of, 66; area of high but declining per capita, 57-61, *charts*, 58, 59; area of increasing per capita vote, 61-66, *charts*, 62, 63; area of moderate but consistent, 67-70, *chart*, 67; areas with little or no, 71-73, *chart*, 72; ratio of farm to non-farm prices and voting, 1870–1948, 83, *chart*, 84; for head of tickets in elections, 1872–1948, *tab.*, 173 f.; geographic patterns of protest, in Presidential elections, 50-77 *with charts;* incidence of poll tax on, 132; potential strength of organized labor rarely available to, 94
Farmer-Labor party, 1920–32, 9; acceptance of La Follette candidacy, 105; agrarian socialism, 113; distribution of votes cast for, 52; La Follette supported by, 23
Farmer-Labor party, Minnesota, *see* Minnesota Farmer-Labor party
Farmer parties, *see* Farmer-labor parties
Farmers, attitude re labor's fight for higher pay, 151; and re labor unions and big business, 152; influence of commercial and subsistence farmers on politics, 148; interchange between field and factory workers, 149; interests and ideologies of labor and, 144, 146, 148; little recent interest in protest parties, 170; operator group politically akin to executive class and proprietors, 161; prices and farm distress, 82-88, 99 (*see also* Prices); reaction to freight charges, taxes, and interest payments, 88
Farmers' Non-Partisan League, 9, 10, 159; still alive in North Dakota, 75; *see also* Non-Partisan League
Farm loan banks, 21
Farm prices, *see* Prices
Farm Relief and Inflation Act, 85
Farm tenancy, *see* Tenant farmers
Federal courts, use of injunctions in labor disputes, 20
Federal Deposit Insurance Corporation, 26
Federal Farm Board, 21
Federal Housing Administration, 26
Federal Reserve Act of 1914, 21
Field, James G., 110
Filing fees required from candidates, 127
Finance, problems of party organization and, 123
Fission, differing effects on minor and major parties, 108
Ford, Ebenezer, 108
Foreign-born, radicalism among, 66
Foreign-policy questions introduced into elections, 76
Fortune Survey, 137, 139, 159, 163, 165 f.
Foster, William Z., 113
Foulke, William D., 16
Freedom of speech and of press, 23
Freese, Irving, 106
Freight rates, 87, 88
Fusion, menace of, 110-12, 123
Fusion candidates, voting for, after merger of Democrats and Populists, 1896, 80
Fusion faction of People's party, 20

INDEX

George, Henry, 7
Goldstein, Jonah J., 120
Gompers, Samuel, 8, 106
Government, control of, by private monopoly, 28; provision for the needy, 143
Government intervention into economic affairs, 138 ff.; groups sympathetic to increased, 30, 153, 170
Government ownership, 21, 22, 23, 27, 139 ff., 153, *tabs.*, 141, 145; of means of production, 139 ff.; polls on, *with tabs.*, 140 ff.
Grain elevator charges, regulation, 14
Granger cases, 14
Granger movement, 13, 41, 42, 117, 168, 169; challenge to Bourbon Democracy, 130, 131; railroad regulation grew out of, 87
Grant, Ulysses S., 16
Great Lakes region, 53, protest voting, 61, 64, 74; strength of farmer-labor parties, 169, 170
Green v. Frazier, 26
Greenback movement, 14-18, 38
Greenback [Labor] party or National party, 7, 16, 17, 103, 168, 169, influence, 104; votes given, 32, 49, 51, *chart,* 63, 65, 102; vote for Butler, 37; geographic pattern of votes for, 73; vote for, affected by price break, 78; attack upon major parties, 96

Harvey, William "Coin," 113
Hayes, Rutherford B., 16, 130
Hillman, Sidney, 110
Home Building Association, state-owned, 25
Home Building Association Act, 26
Hoover, Herbert, 114
Hours of labor, eight-hour day, 15; reduction of, demanded, 17; strike for ten-hour day, 12
Housing, government-sponsored, 29, 142, 144
Hummel, Edward J., 125*n*
Humphrey, Hubert H., 40

Ideas and issues, attitudes of farmers and labor compared, 144 ff.; farmer-labor parties as popularizers of, 168, 172
Imprisonment for debt, 13
Income limitation, opinion poll, 150 f.; *see also* Wealth, redistribution
Income tax, 16, 21, 22, 139
Independent party of 1874, 7; *see further* Greenback party
Independent Progressive party in California, 123
Industrial Commission Act, 26
Industry, control of, by private monopoly, 28; geographic concentration, 64; nationalization, 139 ff., 153, *tabs.*, 141, 145; regulation during war, 147; *see also* Business
Inflation, Greenback demand for, 14; favored by Labor Reform party of 1872, 15
Inheritance tax, 21
Initiative and referendum, 20, 21, 22, 24
Injunctions, use of, by federal courts, 20, 22, 24
Insurance, unemployment, accident, and old-age, 22
Insurance companies, nationalization, 141, 144, *tabs.*, 141, 145
Interest payments on mortgage debt and farmers' agitation, 86, 88
Intermediate credit banks, 21
International Ladies Garment Workers, 109, 122
Interstate Commerce Act of 1887, 14, 15

Jacobstein, Congressman, 116
Jobless party, 113
Judicial review by Supreme Court, 23

Knights of Labor, 18

Labor, attitudes re prosperity and power of farm group, 152, 153;

Labor (*Continued*)
desirability of political action by, 158; deterioration of bargaining position, 1920, 92; endorsement of La Follette, 93; factors that drove toward political action, 93; interests and ideologies of farmers and, 144, 146, 148; use of injunctions in disputes, 20, 22, 24
Labor bureau, establishment of demanded, 17
Labor parties, attitude of workers toward idea of a national party, 158 ff.; early, 6, 12; Roosevelt supported by, in 1944, 74; state and city, 7; vote of all combined, 102; *see also* Farmer and labor parties
Labor Reform party, 6, 15; distribution of votes cast for, 51
Labor's League for Political Education, 10, 107
Labor's Non-Partisan League, 9, 159
Labor unions, activity and voting, 92-95, 97; early parties and, 12; membership, 94; membership attitudes toward politics, 162; period of consolidation, 97; political outlet sought in New York City, 122; public opinion on role in politics, 158 ff.; pure and simple unionism, 106-8; Roosevelt reform legislation a boost to, 93
Labour party, British, 171
La Follette, Philip, 26, 155
La Follette, Robert Marion, effort to combine farmers and workers in concerted political action, 50; endorsed by AFL, 93, 105; lack of a functioning organization, 116; party fell apart after death of, 48; personal drawing power, 97; Presidential contest, 8, 9, 23 f., 104; Socialist backing, 65, 105; success in Western and agricultural states, 40, 83; votes cast for, 33, 36, 48, 52, 57, 70, 74, 82, 97; *see also* Progressive party of 1924
La Follette, Robert M., Jr., 26, 40
La Guardia, Fiorello H., 41, 119, 121

Laws, public: many traceable to agitation by farmer and labor parties, 12, 23, 29
Legal barriers placed in path of minority parties, 125-36
Lehman, Herbert, 40, 118, 121
Lemke, William, 52, 71
Lewinson, Paul, 129
Liberal party of New York State, 53, 169; and ALP relationship, 110, 120, 122; candidates endorsed by, 74, 118 f., 120; formed, 9; membership, 121; record at polls, 33 ff. *passim*, 40, 66, *chart,* 63
Liberal Republicans, 3, 46, 108
Liberty party, votes polled, 1932, 113
Liquor traffic, 17
Loans, borrowing against farm products, 21
Loco-Focos (Equal Rights party), 6
Long, Huey, 70
Lundberg, George A., 65

McKinley, William, 96
Macmahon, Arthur, quoted, 116
Macune C. W., President of Southern Alliance, 18
Mahone, William, 73
Manufacturing, *see* Factories
Marcantonio, Vito, 120
Marketing, efforts to correct injustices in, 88
Marketing corporation, federal, 23
Marxist parties, 45, 47
Massachusetts Labor Reform party, 6
Maurer, James H., 116
Mead, James M., 41, 120
Merchant marine, government ownership of, 28
Middle-of-the-Road faction of People's party, 20
Migrants, disfranchisement, 134
Mill and Elevator Association Act, 26
Milwaukee, Wisc., socialist mayor, 106
Miners, opinion on wealth limitation, 150; on government control, 154
Mines, inspection of, 17

INDEX

Minnesota, election of state legislators, 135 ff.
Minnesota Democratic-Farmer-Labor party, 123
Minnesota Farmer-Labor party, 9, 25, 26, 53, 65, 66, 74, 75, 169; candidates supported by Democratic party, 39; continuous representation in Congress, 42, 45; factors leading to organization of: ascendancy: failure to capture legislature, 135 f.; in state elections, 39; state domination, 117
Mittelman, E. B., quoted, 93
Molly Maguire riots, 96
Monopolies, control of government and industry by private, 28; denounced, 13, 16, 17, 23, 27; enemy of farmer and labor parties, 30
Mortgage debt, 85; interest payments, 86, 88
Munn v. Illinois, 14

Naftalin, Arthur, 135
National Civil Service Reform League, 16
National Industrial Recovery Act, 93
Nationalization, *see* Government control
National Labor party, 104
National Labor Union, objectives, 104; promoted cause of Greenbackism, 15, 29
National Molders' Union, 104
National Opinion Research Council, 167
National party, *see* Greenback Labor party
National Progressives of America, 26
"Nationals" of 46th Congress, 46
Negroes, effect upon votes, 69; voting by, 129, 130; would benefit under two-party system in South, 134
New Deal, *see under* Roosevelt, Franklin D.
New Republic, 125n
Newspapers, rarely support farmer-labor parties, 115, 126

New York City, mayorality elections, 119, 120, 121; protest vote, 53; underrepresented in state legislature, 43
New York Civil Service Reform Association, 16
New York County, 1948 contest for Surrogate, 120
N.Y. State, balance-of-power technique, 117 ff.; electoral law re nomination of candidate, 118; leftist politics, 121; protest vote, 53
Nominating petition, 129
Non-Partisan League, 25, 39, 55, 105, 135; domination of North Dakota by, 117; Republican party in North Dakota controlled by, 25, 74; vote used as yardstick of political radicalism, 65
Non-partisan organizations, 10; *see also titles, e.g.,* Farmers' Non-Partisan League; Labor's . . . ; etc.
Non-partisan pressure groups, reform popularized by, 29
Norris-La Guardia Act of 1932, 20
North Dakota, political background, 54 ff.
Northern Alliance, 18
Northwest, radicalism and conservatism in, 65
Norwalk, Conn., Socialist mayor, 106

O'Dwyer, William, 41, 119, 120
Omaha Resolutions of the Populists, 19
One-party system in South, 129 ff., 133
Open shop, 98
Opinion, influence of education in formation of, 166; lack of, increases in proportion to decline in economic status, 165; re role of unions in politics, 159
Opinion polls, 137 ff.; *tabs.,* 141, 145
Organization, problems of political, 114-17, 123
Owen, Robert Dale, 13, 109

INDEX

PAC, *see* Congress of Industrial Organizations, Political Action Committee

Panics, *see* Depressions

Paper tender issue, 16

Parties, political: absorption of protest vote by major, has prevented consolidation of third parties, 80; class lines shunned by most Americans, 165; control achieved in three ways, 117; election ballot, 125; flexibility, 114, 123, 169; fissions, 108; legal barriers placed in path of minority, 125-36, 171; members, potential reservoir, 172; national farmer-labor party an alternative to realignment of major, 171; national party may be organized by labor, 170; proposal for realignment of major parties, 155 ff.; radical or extremist, have not appealed to electorate, 45, 49, 169; survey, 6-11; *see also under names of parties e.g.,* Democratic party

Paupers, *see* Poor

Pendleton Act of 1883, 16

People's party, *see* Populists, or People's party

Petitions, signatures required, 126; nominating, 129

Platforms, 12-31

Political action, class mobilization for, 102; labor's moves toward, 93, 159

Political Action Committee of the CIO, *see under* Congress of Industrial Organizations

Political allegiance, similarities in, between fathers and children, 165

Political awareness, lack of, among less privileged groups, 165, 166

Political discontent, *see* Discontent

Political expression, restrictive legislation denies freedom of, 128

Political insurgency in Western agricultural states, 170

Political parties, *see* Parties

Political protest, *see* Protest

Politics, balance of power type, 64; elections, 44; recurrent pattern, 30

Polls, *see* Elections; Opinion polls

Poll tax, 68, 131 ff.

Poor, effort to disfranchise paupers, 134; government provision for, 143

Poor whites, limitations on voting, 129, 131, 133

Poor Man's party, 109

Population shifts, 50 ff.

Populists, or People's party, 7, 18-21, 33, 36, 41, 42, 117, 168, 169; battle between Southern Democrats and, 130 ff.; coalition with Republicans, 73, 110, 131; effect of poll tax on, 132; fusion with Democrats, 58 ff., 80, 110 ff.; geographic pattern of votes for, 52, 60, 73; height of power, 38, 96; preamble to 1892 platform, 27; press coverage inadequate, 116; rapid disappearance, 97; record at polls, 1892, 48, 49, 60; studies of, in Georgia and Texas, 69

Postal savings system, 21

Presidential campaigns, *see* Elections, presidential

Press, rarely supports farmer-labor parties, 115, 126

Press, Cleveland, 115

Press, Pittsburgh, 126

Price cycle, peaks in protest voting occur in downswing of, 78

Prices, and farm distress, 82-88, *chart,* 84; question of government regulation, 142; stabilization of farm, 21

―――― wholesale: duration of business cycles dominated by trend of, 91; voting for farmer and labor tickets and, *chart,* 79

Primaries, direct and Presidential preference, 21

Prison labor, abolition of contract system demanded, 17, 21

Production, overexpansion in relation to demand, 85; public ownership of means of, 22 (*see also* Government ownership)

INDEX

Progressive Citizens of America, 117
Progressive party of Illinois, 122
Progressive party of 1912 (Bull Moose party), 3, 10, 108, 116
Progressive party of 1924 (La Follette Progressives), 8, 9, 23 f., 33, 36; attempt to revive in 1938, 155; campaign fund, 115; effect of price drop following postwar boom, 78; impetus and early supporters, 104; La Follette slates on all state ballots except one, 136; nationalization of industry considered, 139; organizational difficulties, 114 ff.; platform, 28; press coverage inadequate, 115; record at polls, 48; *see also* La Follette, Robert M.
Progressive party of 1948 (Wallace movement), 27-29; a farmer-labor party, 10; center of activity, 99; CIO–PAC against, 162; difficulty of getting on ballot, 125n, 126, 127; domestic problems considered, 77; foreign policy issues stressed, 76; groundwork for campaign, 117; inadequate press coverage, 116; on all state ballots except three, 136; platform, 28; symptomatic of political unrest, 170; *see also* Wallace, Henry A.
Progressive party of Wisconsin, *see* Wisconsin Progressive party
Property tax, 16, 87; farmers' reaction, 88
Proportional representation, 22
Protest, political: beginning of stirrings on a national scale: centers of the movements, 170
Protest parties, 4 f.; farm leaders show little interest recently, 170; greatest voting potential, 154; strength lost during depressions because of disfranchisement of migrants, 134; support changing from rural and farm to urban and labor, 42; *see also* Farmer and labor parties; Farmers' parties; Labor parties; *and under names of parties, e.g.,* Anti-Monopoly party

Protest voting, absorption by major parties, 80; components: economic and political discontent, 137-54, 155-67; economics of, 5, 48, 68, 77, 78-101, *with charts;* of farm population, again possible, 170; peaks occur in downswing of price cycle, 78; permanence of, 168; shifts in locale, 169; urban, in N.Y. State; for parties of the left, 121
Public opinion, *see* Opinion
Public ownership, *see* Government ownership
Public utilities, government ownership, 23, 27, 29; regulation of rates, 139, *tabs.*, 141, 145
Public works for the unemployed, 22
Purchasing power of dollar and voting, 89

Race, Class, and Party (Lewinson), 129
Race issue in politics, 69, 73, 129 ff.
Radicalism, economic and political, 163; in Northwest, 65
Railroad brotherhoods, 105, 170
Railroads, nationalization, 21, 23, 28, 140, *tabs.*, 141, 145; regulation, 14, 16, 87
Railway and Warehouse Commission, 14
Reading, Pa., socialist mayor, 106
Readjuster movement in Virginia, 72
Recall of elected officials, 21, 22
Reconstruction Acts, 130
Reform groups, effort of farmer and labor parties to form working alliances with local and state, 106; tendency in N.Y. to unite against Tammany, 121
Reform movement, People's party gave impetus to, 20
Reform party, 7
Reforms traceable to labor parties, 12, 29, 155
Relief recipients, attempts to disfranchise, 134
Republican-Liberal coalition, 41

Republican party, AFL move directed against, 107; campaign fund, 1924, 115; factionalism, 108; Non-Partisan League's control of, in North Dakota, 25, 74; secession of Progressives of 1912 from, 10; support of "farm bloc" in Congress, 60
Republican-Populist coalition, 73, 110, 131
Residential requirements for voting, 129, 134
Reuther, Walter, 170
Reynolds, Verne L., 113
Roosevelt, Franklin D., 9; elected on program designed to aid farm and labor population, 85; electoral discontent relatively stabilized during elections for, 36; groups supporting, 35, 39, 74; income taxes used to finance social reform of administration, 139; party absorbed discontent following economic collapse, 68; public reaction to political system during administration of, 156, 157; reform legislation a boost to organized labor, 93; Supreme Court reform, 24; votes for, 114, 118 f., 121; Wisconsin Progressives cooperated with, 26
—— New Deal, 23, 26; failed to bring permanent recovery, 138; farmer-labor support of program, 98, 99, 170; opinion polls re measures started, 153; polarization of interests within two-party system during era of, 171; steps to aid unemployed, 139
Roosevelt, Franklin D., Jr., 41
Roosevelt, Theodore, 10, 116
Russia, Progressives called for cooperation with, 28

St. Louis convention of 1889, 18
Sanitary laws, 17
Schattschneider, E. E., 5
Schools, see Education
Schurz, Carl, 16

Scripps-Howard newspapers, 115, 126
Sewall, Arthur, 111
Shipstead, Henrik, 39, 135
Signatures on petitions, legal requirements, 126
Silver issue, 19, 20, 111, 113
Silver Republicans, 3, 108
Single-Taxers, 31
Skidmore, Thomas, 12, 109
Smith, Gerald L. K., 9
Social conditions, influence on vote for farmer-labor parties, 48
Social-Democratic party of America, 8, 22
Socialism, agrarian, 25
Socialist Labor party, 33, 47; founders, 7 f.; in campaigns of 1892, 8, 21; of 1932, 113; of 1944, 8; Marxism of, 7 ff.; platform of 1892, 21
Socialist mayors, 106
Socialist movement, 21-23
Socialist Party of America, beginnings, 8, 22; carried principal banner of discontent, 65; Debs its grand old man, 8; foreign-born in strongholds of, 66; geographic pattern of vote in 1912, 74; La Follette candidacy accepted, 105; long record of, 47; N.Y. State vote, 74, 121; 1932 campaign, 112, 113, 162; only minor party with a relatively fixed ideology to enjoy success, 46; peak of strength in 1912, 36, 74, 91, 97; representation in Congress, 45 f.; shift of electoral support, 52; vehicle for electoral discontent in South, 68, 169; votes polled, 1920, 33
Social security, 27, 29
Social Security Act of 1935, 22
"Sons of the wild jackass," 60
South, battle between Democrats and Populists, 130 ff.; limitations on voting: one-party system, 129 ff., 133; low level of voter participation, 133; probable return to, and effects of, two-party system, 134; shift in protest area, 51; Socialist

INDEX

party the vehicle for electoral discontent, 68
South Dakota, political background, 54 ff.
Southern Alliance, 18, 130
Southern Democrats, *see* Democrats, Southern
Southern Populism, breakdown of, 110 ff.
Specie payments, partial suspension of, 1873, 96
Specie Resumption Act, 16
Stassen, Harold, 25, 39
State elections, *see* Elections, state
State legislatures, underrepresentation in, 42
State parties, recent, 24-27
States, having poll tax requirement, 131 ff.; tax repealed, 132; with laws restricting Communist and other parties, 127; *see also* Elections, state
Stock market collapse, *see* Depressions
Streeter, A. J., 48
Strikes, and protest voting, 93; symptoms of economic discontent, 94; trend in, 92
"Sub-treasury" scheme, Macune's, 19
Suffrage, poll tax and, 68, 131 ff.; residential requirements for, 129, 134; woman, 17, 20
Surpluses, purchasing of, 21
Sylvis, William, 103

Taft-Hartley Act of 1947, 20, 24, 29; AFL move to combat, 107
Talmadge, Eugene and Herman, 70
Tammany challenged by Equal Rights party, 6
Taxation, *see* Income tax; Inheritance tax; Poll tax; Property tax
Taylor, Glen, 10
Tenant farmers in South, 69; political alignment, 73
Third parties, 3-5; two main categories, 3 f.; *see also* Farmer-labor parties; Protest parties; *also their names, e.g.,* Socialist party

Thomas, Norman, 113
Thorp, W. L., quoted, 91
Tillman, Ben, 21
Tillman Democrats, 130
Times, New York, 120
Townley, A. C., 25, 55
Townsend, Francis, 9
Transients, disfranchisement, 134
Transportation charges, 87
Trotskyites, 31
Truman, Harry S., California carried by, 123; Liberal party endorsement of, 119; protest vote supported, 100; Wallace supporters go over to, 49
Two-party system, 3; adequate to meet challenges of depression and war, 171; bulk of citizenry subscribe to, 164; polarization of interests during New Deal era, 171; Southern states, 130, 134

Unemployment, during the depression of 1930's, 138; government responsibility for alleviating, 138, 140, 144 ff., *tabs.,* 141, 145
Union Labor party, 7, 46, 48; distribution of votes cast for, 51; platform of 1888, 18
Union party of 1936, 9, 71; votes cast for, 52, 162
Unions, *see* Labor unions
United Automobile Workers, 170
United Labor party, 7
United States, Congress: dissatisfaction with Senate, 18; direct election of Senators, 18, 19, 20, 22; "farm bloc," 60; parties and states of farmer and labor party representatives in, 1860-1948, *list,* 175-77; the eightieth, 99
—— Constitution: Amendments, 17-20 *passim*
—— Department of Labor: Bureau of Labor Statistics, 22
—— Supreme Court: Granger cases, 14; reform proposed, 24
Urbanization, trend toward, 50 ff.
Utilities, *see* Public utilities

190 INDEX

Virginia Readjusters, 72
Voting, see Elections; Protest voting

Wages and voting, 88-89
Wagner Act, 24, 93
Waite, Chief Justice, 14
Wallace, Henry A., 9, 10, 70; ALP endorsement of, 108, 110, 119; distribution of votes cast for, 53; failure to qualify in Illinois, 127; obstacles in way of sponsors, 125n; protest against Democratic party's policies, 85; record at polls, 35; speech at Chicago, Dec. 29, 1947, 28; supporters go over to Truman, 49; support in California, 123
Wallace movement, see Progressive party of 1948
Warehouse Act, 21
Warehouse charges, demands for state regulation, 14
Washington, state, political background, 56
Washington Commonwealth Federation, 56
Watson, Thomas, 111

Wealth, redistribution, 139, 144, 151, 170, tabs., 141, 145
Weaver, James B., 32, 33, 36, 48, 70, 110; distribution of votes cast for, 51, 57
Webb, Frank E., 113
Welfare legislation, 27
West, shift in protest vote, 51; strength of farmer-labor parties in, 169, 170
Wheeler, Burton K., 23, 115
Wheeler, Everett P., 16
Willkie, Wendell, 118
Wisconsin Progressive party, 9, 26, 53, 65, 66, 74, 169; continuous representation in Congress, 45; F. D. Roosevelt supported by, 39; spectacular role, 40, 42; state domination by, 117
Woman suffrage, 17, 20
Women in industry, 21
Workingmen's parties, 1827-38, 6, 78, 108
Working People's Political League, 25
Wright, Frances, 13, 109
Write-in vote, 129

JAN 11

3 3311 00289840 3